Lecture Notes in Business Information Processing 491

LNBIP reports state-of-the-art results in areas related to business information systems and industrial application software development – timely, at a high level, and in both printed and electronic form.

The type of material published includes

- Proceedings (published in time for the respective event)
- Postproceedings (consisting of thoroughly revised and/or extended final papers)
- Other edited monographs (such as, for example, project reports or invited volumes)
- Tutorials (coherently integrated collections of lectures given at advanced courses, seminars, schools, etc.)
- Award-winning or exceptional theses

LNBIP is abstracted/indexed in DBLP, EI and Scopus. LNBIP volumes are also submitted for the inclusion in ISI Proceedings.

Julius Köpke · Orlenys López-Pintado ·
Ralf Plattfaut · Jana-Rebecca Rehse ·
Katarzyna Gdowska ·
Fernanda Gonzalez-Lopez · Jorge Munoz-Gama ·
Koen Smit · Jan Martijn E. M. van der Werf
Editors

Business Process Management

Blockchain, Robotic Process Automation and Educators Forum

BPM 2023 Blockchain, RPA and Educators Forum
Utrecht, The Netherlands, September 11–15, 2023
Proceedings

 Springer

Editors
Julius Köpke 🆔
University of Klagenfurt
Klagenfurt, Austria

Ralf Plattfaut 🆔
University of Duisburg-Essen
Essen, Germany

Katarzyna Gdowska 🆔
AGH University of Science and Technology
Krakow, Poland

Jorge Munoz-Gama 🆔
Pontificia Universidad Católica de Chile
Macul, Chile

Jan Martijn E. M. van der Werf 🆔
Utrecht University
Utrecht, Utrecht, The Netherlands

Orlenys López-Pintado 🆔
University of Tartu
Tartu, Estonia

Jana-Rebecca Rehse 🆔
University of Mannheim
Mannheim, Baden-Württemberg, Germany

Fernanda Gonzalez-Lopez 🆔
Pontificia Universidad Católica de Chile
Santiago, Chile

Koen Smit 🆔
University of Applied Sciences Utrecht
Utrecht, The Netherlands

ISSN 1865-1348   ISSN 1865-1356 (electronic)
Lecture Notes in Business Information Processing
ISBN 978-3-031-43432-7   ISBN 978-3-031-43433-4 (eBook)
https://doi.org/10.1007/978-3-031-43433-4

This Springer imprint is published by the registered company Springer Nature Switzerland AG
The registered company address is: Gewerbestrasse 11, 6330 Cham, Switzerland

Paper in this product is recyclable.

Preface

In 2003, the International Conference on Business Process Management (BPM) started as the conference where people from academia and industry can meet to discuss the latest advances in the field. In 2023, the conference was held in Utrecht, The Netherlands. The conference has a tradition to feature forums to discuss specialized topics. This volume contains the proceedings of the Blockchain Forum, the Robotic Process Automation (RPA) Forum, and the Educators Forum, which took place during September 11–15, 2023.

Blockchain systems offer desirable features like immutability, integrity, and distributed consensus to support cross-organizational information systems and, in particular, Inter-Organizational Business Processes. Indeed, many research opportunities exist across all phases of the BPM lifecycle. Blockchain-based applications have proved robust capabilities across various domains, including logistics, healthcare, and finance. However, limitations like privacy and scalability still present significant challenges for many practical applications. In this setting, the fifth edition of the Blockchain Forum provided a platform for discussing and showcasing ongoing research and applications at the intersection of BPM and blockchains.

Robotic Process Automation (RPA) is a maturing technology in the field of Business Process Management. It allows the development of (multiple) computer programs (i.e., bots) that automate rules-based business processes through the use of GUIs. As such, it enables the office automation of repetitive tasks. The technology itself has reached a certain level of technological maturity and organizational adoption. Hence, researchers now have the chance to look at RPA in a larger context, for example with regard to low-code automation, RPA for smart automation, or organizational implications and considerations of RPA. The fourth edition of the RPA forum at the 21st International Conference on Business Process Management aimed to provide a forum for a broad range of researchers to exchange ideas, fuel new research, and start corresponding collaborations.

The Educators Forum seeks to bring together educators within the BPM community to share resources to improve the practice of teaching BPM-related topics by exchanging experiences, bringing in cases, and discussing teaching innovations. The Educators Forum was held for the first time, and offered a meeting point for lecturers to share, discuss, and improve our BPM teaching in a dynamic way, far from more traditional research contributions.

This year, the blockchain forum attracted eight submissions, of which four were finally selected for publication and presentation. The RPA forum received 21 submissions, of which eight were selected for publication and presentation. The Educators Forum had 13 submissions, of which six were selected for publication and presentation. This results in an overall acceptance rate of 43%. For each of the forums, the papers were peer-reviewed by three members of the respective program committees in a single-blind review process.

As forum organizers, we hope that the reader will enjoy the papers selected for the different forums, and that the presentations and content will lead to interesting discussions. We want to thank all authors who submitted a paper to the forum. Unfortunately not all papers were selected, but they were all of high quality, and the committee enjoyed reading them. We want to express our gratitude to everyone who served in the review process for their critical, fruitful comments. A last word of thanks to Xixi Lu and Felix Mannhardt, who helped in compiling this volume.

September 2023

Julius Köpke
Orlenys López-Pintado
Ralf Plattfaut
Jana-Rebecca Rehse
Katarzyna Gdowska
Fernanda Gonzalez-Lopez (Deceased)
Jorge Munoz-Gama
Koen Smit
Jan Martijn E. M. van der Werf

Organization

Blockchain Program Chairs

Julius Köpke · · · · · · · · · · · · · · · · University of Klagenfurt, Austria
Orlenys López-Pintado · · · · · · · · · · University of Tartu, Estonia

RPA Program Chairs

Ralf Plattfaut · · · · · · · · · · · · · · · · University of Duisburg-Essen, Germany
Jana-Rebecca Rehse · · · · · · · · · · · · University of Mannheim, Germany

Educators Program Chairs

Katarzyna Gdowska · · · · · · · · · · · · AGH University of Science and Technology,
    Poland
Fernanda Gonzalez-Lopez · · · · · · · · Pontificia Universidad Catòlica de Chile, Chile
  (Deceased)
Jorge Munoz-Gama · · · · · · · · · · · · Pontificia Universidad Catòlica de Chile, Chile
Koen Smit · · · · · · · · · · · · · · · · · · University of Applied Sciences Utrecht,
    The Netherlands
Jan Martijn E. M. van der Werf · · · · Universiteit Utrecht, The Netherlands

Combined Program Committee

Claudio Di Ciccio · · · · · · · · · · · · · Sapienza University of Rome, Italy
Marco Comuzzi · · · · · · · · · · · · · · · Ulsan National Institute of Science and
    Technology, South Korea
Walid Gaaloul · · · · · · · · · · · · · · · · Télécom SudParis, France
José María García · · · · · · · · · · · · · University of Seville, Spain
Felix Härer · · · · · · · · · · · · · · · · · · University of Fribourg, Switzerland
Marko Hölbl · · · · · · · · · · · · · · · · · University of Maribor, Slovenia
Kais Klai · Sorbonne Paris Nord University, France
Qinghua Lu · · · · · · · · · · · · · · · · · · CSIRO, Australia
Raimundas Matulevicius · · · · · · · · · University of Tartu, Estonia
Giovanni Meroni · · · · · · · · · · · · · · Technical University of Denmark, Denmark

Andrea Morichetta	University of Camerino, Italy
Alex Norta	Tallinn University of Technology, Estonia
Pierluigi Plebani	Politecnico di Milano, Italy
Stefan Schulte	TU Hamburg, Germany
Tijs Slaats	University of Copenhagen, Denmark
Mark Staples	CSIRO, Australia
Horst Treiblmaier	MODUL University Vienna, Austria
Ingo Weber	Technical University of Munich, Germany
Kaiwen Zhang	École de technologie supérieure, Canada
Simone Agostinelli	Sapienza University of Rome, Italy
Aleksandre Asatiani	University of Gothenburg, Sweden
Bernhard Axmann	Technical University of Ingolstadt, Germany
Adela Del Río Ortega	University of Seville, Spain
Carmelo Del Valle	University of Seville, Spain
Mathias Eggert	FH Aachen, Germany
Carsten Feldmann	FH Münster University of Applied Sciences, Germany
Peter Fettke	German Research Center for Artificial Intelligence (DFKI), and Saarland University, Germany
Christian Flechsig	Technische Universität Dresden, Germany
Norbert Frick	Hochschule der Deutschen Bundesbank, Germany
José González Enríquez	University of Seville, Spain
Lukas-Valentin Herm	Julius-Maximilians-Universität Würzburg, Germany
Hannu Jaakkola	University of Tampere, Finland
Christian Janiesch	TU Dortmund University, Germany
Andrés Jiménez Ramírez	University of Seville, Spain
Fabrizio Maria Maggi	Free University of Bozen-Bolzano, Italy
Andrea Marrella	Sapienza University of Rome, Italy
Massimo Mecella	Sapienza University of Rome, Italy
Dan O'Leary	University of Southern California, USA
Teijo Peltoniemi	University of Turku, Finland
Yara Rizk	IBM Research, USA
Rehan Syed	Queensland University of Technology, Australia
Maximilian Völker	Hasso Plattner Institute, Germany
Moe Wynn	Queensland University of Technology, Australia
Matthijs Berkhout	University of Applied Sciences Utrecht, The Netherlands
Marlon Dumas	University of Tartu, Estonia
Mahendra Er	Institut Teknologi Sepuluh Nopember, Indonesia
Daniele Grigori	Laboratoire LAMSADE, University Paris-Dauphine, France

Sam Leewis   University of Applied Sciences Utrecht,
                            The Netherlands
John van Meerten   University of Applied Sciences Utrecht,
                            The Netherlands
Jan Mendling   Humboldt-Universität zu Berlin, Germany
Gregor Polancic   University of Maribor, Slovenia
Pascal Ravesteijn   Utrecht University of Applied Sciences,
                            The Netherlands
Manuel Resinas   University of Seville, Spain
Mojca Stemberger   University of Ljubljana, Slovenia

Additional Reviewers

Davide Basile
Mubashar Iqbal
Yue Liu

Contents

Educators Forum

Blockchain Forum

Preface

The BPM 2023 Blockchain Forum provided a platform for exploring and discussing innovative ideas on the intersection of BPM and blockchain technology. Associated with the BPM conference, this year's Blockchain Forum in Utrecht continued the legacy of prior forums in Münster (2022), Rome (2021), Seville (2020), and Vienna (2019). A growing body of research combining BPM and blockchain/distributed ledgers illustrates blockchain technology's academic and practical appeal in the scope of Business Process Management. Specifically, this year, the forum received eight submissions, of which four were carefully selected based on rigorous reviews for presentation at the Blockchain Forum and inclusion in the proceedings.

The paper "Towards Object-centric Process Mining for Blockchain Applications" addressed the challenges of process mining in blockchain applications, particularly in creating event logs. Hobeck and Weber highlighted issues with the traditional event log format, XES, which only allows for single-case notions, leading to problems like divergence and convergence. The paper proposed the multi-case Object-Centric Event Log (OCEL) format as a solution to challenges with single-case event logs. Also, the authors presented an approach to extracting data from blockchain applications and mapping it to OCEL while broadening the scope of considered data sources, thereby providing a more comprehensive data capture.

Lichtenstein et al. investigated the role of 'looseness', the configuration and execution of underspecified processes, in blockchain-based collaborative procedures in their paper "Loose Collaborations on the Blockchain: Survey and Challenges". The authors presented a systematic literature review to analyze how blockchain-based approaches manage looseness in collaborative processes from both behavioral and organizational perspectives. The paper also identified open challenges in supporting looseness within these processes, opening avenues for further research.

In the paper "ChorSSI: a BPMN-Based Execution Framework for Self-Sovereign Identity Systems on Blockchain", the authors Cippitelli, Marcelletti, and Morichetta introduced a framework to facilitate the creation and operation of Self-Sovereign Identity (SSI) systems, a type of decentralized identity model built on blockchain technology. Operating on BPMN choreography diagrams, ChorSSI represents SSI interactions from a high-level perspective, making the system more accessible to non-expert users. The authors implemented and tested the ChorSSI framework through a real-world case study on Chromaway property transactions.

In the study "Towards an Understanding of Trade-offs Between Blockchain and Alternative Technologies for Inter-Organizational Business Process Enactment," Kjäer, Preindl, and Kastner explored blockchain's role in Inter-organizational Business Processes (IOBPs). They emphasized the lack of blockchain comparisons with conventional, non-blockchain systems. The authors fill this void by contrasting blockchains and technologies like Trusted Third Parties (TTPs) and Electronic Data Interchange (EDI), highlighting blockchain's unique benefits like non-equivocation while acknowledging TTPs' simplicity and flexibility. They advocated for careful analysis of trade-offs when considering blockchain for Business Process Management System (BPMS) integration, aiming to guide software architects and inspire further blockchain research.

We extend our gratitude to everyone involved in making the BPM 2023 Blockchain Forum successful: the authors who diligently submitted their papers, the Program Committee members and additional reviewers for meticulously reviewing each paper, and the speakers who shared their valuable work. We also thank the BPM 2023 chairs and organizers for their support in preparing the Blockchain Forum.

September 2023   Julius Köpke
                                                                Orlenys López-Pintado

Organization

Program Chairs

Julius Köpke   University of Klagenfurt, Austria
Orlenys López-Pintado   University of Tartu, Estonia

Program Committee

Claudio Di Ciccio	Sapienza University of Rome, Italy
Marco Comuzzi	Ulsan National Institute of Science and Technology, South Korea
Walid Gaaloul	Télécom SudParis, France
José María García	University of Seville, Spain
Felix Härer	University of Fribourg, Switzerland
Marko Hölbl	University of Maribor, Slovenia
Kais Klai	Sorbonne Paris Nord University, France
Qinghua Lu	CSIRO, Australia
Raimundas Matulevicius	University of Tartu, Estonia
Giovanni Meroni	Technical University of Denmark, Denmark
Andrea Morichetta	University of Camerino, Italy
Alex Norta	Tallinn University of Technology, Estonia
Pierluigi Plebani	Politecnico di Milano, Italy
Stefan Schulte	TU Hamburg, Germany
Tijs Slaats	University of Copenhagen, Denmark
Mark Staples	CSIRO, Australia
Horst Treiblmaier	MODUL University Vienna, Austria
Ingo Weber	Technical University of Munich, Germany
Kaiwen Zhang	École de technologie supérieure, Canada

Additional Reviewers

Davide Basile
Mubashar Iqbal
Yue Liu

ChorSSI: A BPMN-Based Execution Framework for Self-Sovereign Identity Systems on Blockchain

Tommaso Cippitelli, Alessandro Marcelletti^(✉), and Andrea Morichetta

University of Camerino, Camerino, Italy
tommaso.cippitelli@studenti.unicam.it,
{alessand.marcelletti,andrea.morichetta}@unicam.it

Abstract. The digital age has made identity a crucial aspect of online activities due to our increasing reliance on digital platforms. This has led to the development of different identity management systems, relying on centralised infrastructures but exposed to security vulnerabilities. Self-Sovereign Identity (SSI) is a promising alternative, as it allows individuals to control their personal data and securely share it with others without relying on a central authority. However, developing such systems and executing the related operations is complex and challenging, especially for non-expert users. To simplify the development process, we propose ChorSSI, a BPMN-based framework that supports the modelling of an SSI system and the execution of the related interactions. The design relies on BPMN choreography diagrams, permitting the representation of SSI interactions between parties in a distributed manner. The proposed framework was implemented and tested over the real Chromaway property transaction case study.

Keywords: BPMN · Choreography · SSI · Blockchain · Execution Framework

1 Introduction

In the digital age, the continuous growth of connected services and technologies has led identity to become a critical aspect of online activities [1]. Indeed, we increasingly rely on digital platforms for communication, commerce, and other activities. Therefore the need for secure and reliable methods for identifying people and things has become more important than ever [2]. To address this challenge, different identity management systems were developed to manage digital identities and their interconnection [3]. However, those systems are often centralised [4] and require the involvement of trusted third parties [5]. Indeed, such solutions can lead to security vulnerabilities like data breaches and identity fraud [6]. A promising alternative to traditional identity management systems is certainly Self-Sovereign Identity (SSI). This is a decentralised identity model that provides individuals control over their personal data and allows them to share data

© The Author(s), under exclusive license to Springer Nature Switzerland AG 2023
J. Köpke et al. (Eds.): BPM 2023, LNBIP 491, pp. 5–20, 2023.
https://doi.org/10.1007/978-3-031-43433-4_1

securely and selectively with other parties, without having to rely on a single central authority [7]. This is possible thanks to blockchain, which is the underlying technology on which SSI systems are usually built on top of [8]. The blockchain integration provides a secure and tamper-proof distributed ledger to store and manage identity information [2]. To enable information sharing, SSI relies on Verifiable Credentials (VCs), which are digital representations about an individual, organisation, or thing and they are issued by trusted parties and can be cryptographically verified [9]. Typical operations of SSI systems are related to the issuing, verification and revocation of VCs done by the actors that interact with an SSI system. Another key aspect of SSI is the confidentiality of interactions between actors, which is ensured by the Zero Knowledge Proof (ZKP) cryptographic protocol, allowing to prove the validity of a statement without sharing the underlying information [10]. ZKP enhances user privacy while maintaining the necessary institutional trust for the correctness of digital interactions and represents one of the key benefits of SSI. While the concept of SSI is relatively new, there has been growing interest in the development and usage of SSI systems and applications. However, building such systems can be complex and challenging for developers, that have to learn different concepts and technologies. Furthermore, the execution of SSI-related operations is usually done by end-users or stakeholders that do not have a technical background. All of these aspects represent barriers to the development and adoption of the SSI model [11]. For this reason, there is a need to integrate low-code strategies that facilitate the development process and makes SSI applications more accessible to a wider range of users.

In this work, we propose ChorSSI, a BPMN-based framework that supports the creation of an SSI system and guides the execution of the related operations. In particular, BPMN choreography diagrams are used to represent the SSI interactions as they enable to represent the communication between distributed parties from a high-level perspective. These diagrams provide a comprehensive representation of SSI interactions by determining the overall execution flow based solely on the definition of exchanged messages [12]. In ChorSSI, each step designed in the choreography represents an SSI protocol specifying which parties are involved and the exchanged information (e.g., a set of credentials). This model is then used to guide the execution of the end-user, providing a high-level interface for interacting with the different components without the need for technical knowledge. This is possible thanks to the supporting mechanisms provided by ChorSSI. Indeed, after the first configuration is done by the developer, ChorSSI automatically completes and connects the software components later exploited by the users, allowing them to better focus on the use case rather than the underlying technology.

To show the effectiveness of the ChorSSI approach, a prototype was implemented on the real-world Chromaway property transactions case study.

The rest of the work is organised as follows. Section 2 introduces the main concepts and technologies when developing an SSI system on blockchain. The section also introduced the considered case study. Section 3 described the ChorSSI methodology and the phases for executing SSI interactions starting from a choreography. In Sect. 4 the ChorSSI prototype is described, highlighting the main phases and functionalities implemented in the proposed case study. Section 5 introduces relevant works while Sect. 6 concludes and provides an overview of future directions.

2   Background

This section presents the main concepts and technologies on which ChorSSI relies on. In particular, a first introduction to SSI systems is given, introducing the actors, components and possible operations related to digital identities. Then, the used Hyperledger technologies are presented with a final description of the used case study.

2.1   Self-Sovereign Identity Model

In an SSI system, individuals create and control their digital identities, which are stored on the blockchain thanks to the use of different concepts and technologies. [13]

SSI Components are the fundamental concepts that characterise self-sovereign identity systems and the element required to comprehend and use an SSI infrastructure.

Verifiable Credential is a digitally signed data structure that contains a set of information about a subject that can be cryptographically verified. A VC is stored on the subject's device or data store and can be shared with others when needed. Overall, Verifiable Credentials enable secure, privacy-preserving, and decentralised management of digital identities and personal data, and are a crucial building block of the SSI model.

Decentralised Identifiers (DIDs) are globally unique identifiers that are cryptographically verifiable and self-administered. DIDs are a critical component of the SSI model, enabling secure, decentralised, and interoperable management of digital identities and personal data. DIDs can be used to represent any entity that requires a unique identifier, such as individuals, organisations, things, or even abstract concepts. The DID is associated with a public-private key pair, for verifying the associated digital signatures.

Wallet is a software application that enables an individual or an organisation to store, manage, and control their digital identities and VCs in a secure and privacy-preserving manner. A wallet includes the ability to receive and store VCs, manage their presentation and access digital identity data.

Zero-knowledge Proof is a cryptographic algorithm that enables the sharing of VCs or other sensitive information without revealing the actual content. Indeed, ZKP allows an entity to create digital proof about the ownership of a secret without actually revealing its value. ZKP algorithms are crucial in SSI systems, especially in situations where the sharing of information needs to be privacy-preserving, such as in healthcare or financial contexts.

SSI Flow is managed through a defined set of protocols and components. In this context, four key entities can be found: Holders, Issuers, Verifiers, and Verifiable Data Registries (i.e., the blockchain) acting according to the following operations.

Issue Credential is the protocol that permits parties to exchange VCs between an Issuer and a Holder and it consists of a sequence of steps. The first one is to create a *credential definition*, which is a template that represents a given type of credential. The Issuer defines the credential schema, including the attributes to be considered (e.g., name, birth date, etc.) and the rules for their verification once the credential is emitted. The second step is the *credential offer*, where the Issuer sends an offer to the Holder, indicating the willingness to issue a credential. The offer includes the schema and any specific attribute values to include. The Holder can then choose to accept or reject the offer. In case of acceptance, the Holder sends a credential request to the Issuer through the *credential request* step. This includes any additional information needed to complete the credential, such as proof of identity. Finally, during the *issue credential*, the Issuer fills the credential, signing it with its private key and sending it to the Holder.

Request Proof is the protocol for requesting and receiving proofs from a Verifier in a secure, and privacy-preserving manner. The first step is optional and corresponds to *propose presentation*, where the Prover sends a proof proposal message to the Verifier. After this, in the *request proof* the Verifier describes what attributes or credentials have to be verified. The message contains the details of the proof request, including the requested attributes or credentials. The message also includes a nonce, which is a random number to ensure the uniqueness of the request. If the Prover decides to fulfil the received request, during the *proof presentation*, the proof is constructed and a confirmation message is sent back. This confirmation message includes both the requested proof and the proof presentation, which serves to demonstrate the validity of the exchanged proof. Subsequently, the Verifier can proceed to verify the proof presentation to ensure its authenticity and confirm that it fulfils the original proof request. If the proof is deemed valid, the Verifier can then use it for the intended purpose.

Revoke of a credential is the last protocol and is performed by an Issuer. In this case, the procedure is rather simple and consists of changing the reference of a valid certificate and its attributes. As a result, when a proof request is invoked, the credential no longer appears valid and is considered revoked.

2.2 Hyperledger Aries and Indy

Hyperledger Aries is an open-source toolkit within the Hyperledger ecosystem that provides a set of protocols and tools for creating, transmitting, and storing verifiable digital credentials. **Aries Agents** are software components that enable secure communication and data exchange between different entities in a decentralised system. These agents use cryptographic protocols and digital credentials to establish trust between parties and facilitate secure data sharing. Aries Agents can be categorised and configured based on the role they play in a decentralised system, including Issuer Agents, Holder Agents, and Verifier Agents.

Hyperledger Indy is a decentralised, open-source blockchain platform that provides the infrastructure for building and using DID solutions. It is specifically designed to address the unique requirements of decentralised identity systems, such as privacy, security, and interoperability. Hyperledger Indy provides several key features that are essential for decentralised identity systems, including the ability to create and manage DIDs, the ability to issue and verify verifiable credentials, and the ability to build privacy-preserving and secure interactions between entities.

2.3   SSI Case Study

This section introduces the ChromaWay property transactions project reported in Fig. 1, a blockchain-based system for managing property transactions in Sweden. This project is a significant example of self-sovereign identity that has been highlighted also in [14] by the European Commission. This use case serves as an inspiration for exploring the potential of ChorSSI in real-world applications. Indeed, the real estate industry involves high-value transactions, making it crucial to ensure the security and transparency of property transfers. However, the current system for real estate transactions is slow, expensive, and prone to errors, including contested property deeds. To address these challenges, the Chromaway project was initiated in September 2016. The goal was to redefine real estate transactions and mortgage deeds, addressing the pain points of the current system, which include the lack of transparency, a slow registration system, and the complexity of agreements between buyers and sellers due to a lack of trust in the system and high transaction costs. The solution introduces a new blockchain-based workflow that streamlines and secures the process of transferring property titles. The Swedish Land Registry stores land titles while the blockchain stores the state of the system after each step in the workflow. The synchronisation of participants involved in the transaction is guaranteed in this way.

3   ChorSSI Framework

This section explains the ChorSSI framework reported in Fig. 2. In particular, we describe the different phases that permit the design of a choreography model representing SSI operations until its final execution and monitoring. In general, ChorSSI exploits choreography diagrams to enable and facilitate the execution of SSI interactions abstracting from low-level and technical details. ChorSSI aims at adding a new layer of abstraction for organisations, allowing them to communicate on top of an existing infrastructure without the need for any technical knowledge. In particular, a choreography represents the execution flow of the implemented SSI system and it permits organisations to specify at run-time all the information needed to communicate in a proper way. Notice, ChorSSI binds choreography messages to the standard SSI protocols which operations

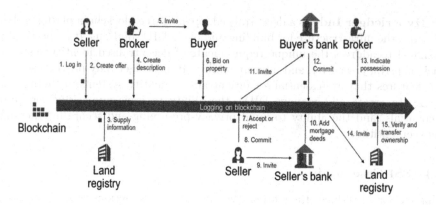

Fig. 1. Chromaway case study.

are already built-in without any translation mechanism. Thanks to this abstraction, during the execution users need only to insert the required information and ChorSSI automatically invokes the underlying infrastructure.

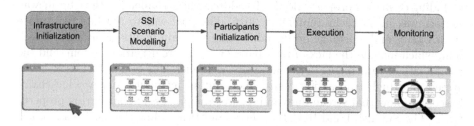

Fig. 2. ChorSSI framework and phases.

Infrastructure Initialisation. The first phase consists of preparing the necessary configuration for the successful launch of the ChorSSI infrastructure. Indeed, current SSI technologies are still at an early stage and they require many technical details to be configured in order to be executed properly. This involves a series of operations aimed at setting up the public ledger acting as the Verifiable Data Registry storing transactions. Additionally, the agents representing the parties of the business process are identified and configured. Finally, the ChorSSI web application is established. This application provides the necessary user interface and functionality for interacting with the public ledger and the agents.

Modelling Phase. The goal of this phase is to create the BPMN choreography diagram representing the different SSI operations. The created model is used to guide the end-user during the execution of the system, abstracting from low-level details and providing the necessary information for completing each task. In particular, ChorSSI supports the **issuing** and **request proof** procedures between participants modelled according to the BPMN standard. The **revoke** instead, does not require to be modelled as it can be invoked at every moment. Such protocols represent standard operations of an SSI system which are built-in inside ChorSSI and they are specified inside choreography messages. The organisations instead, are represented as participants and they are bonded to SSI agents previously configured.

Participants Initialisation. Once the model is designed, it is used as the input for initialising the participants' connections. This is done automatically and it first connects all the participants' agents and then creates the corresponding credential definitions. All these technicalities are facilitated by ChorSSI, which thanks to the choreography and the initial configuration, is able to provide the capabilities needed without requiring additional effort from the user. In particular, at this stage, ChorSSI binds each organisation, represented as choreography participants, to SSI agents exploiting the initial infrastructure configuration. In this way, it is not necessary to code such components.

Execution. During the execution, the participants perform the operations related to the SSI protocols following the choreography model. Indeed, the model represents the execution flow of the implemented SSI system and it permits organisations to specify only the information needed at run-time. To this purpose, ChorSSI provides all the required functionalities through a graphic interface attached to the choreography. Thanks to this, the invocation and communication with the agents and the blockchain are automatically managed. Each party can so concentrate on executing a certain task by simply providing the required information. In this way, ChorSSI reduces the amount of manual coding needed and increases productivity. This also assists the final users to better focus on the complexity of the business process, reducing the time required to manually develop and implement the SSI-based use case.

Monitoring. The last phase enables the participants to observe the executed operations and monitor the current state of the process. This is done by controlling all the produced and exchanged data. In that way, each party is able to check whenever a certain object meets the initial expectations. The monitor is important for several reasons. First, it permits checking that the executed operations meet the requirements and specifications outlined in the starting choreography. Second, it provides an opportunity to test and evaluate the SSI scenario, as it can uncover issues or opportunities for improvement that may not have been identified during the development process. Finally, it provides feedback that can be used to further refine and improve the SSI scenario and the treated use case over time.

4   ChorSSI Prototype

This section shows the prototypical implementation of the ChorSSI framework[1] over the Chromaway case study presented in Sect. 2.3. In particular, the

---

[1] Tool available at https://bitbucket.org/proslabteam/chorssi/src/main/.

scenario was adapted as a choreography reported in Fig. 3. Here the Seller starts by getting the ownership certificate of its property from the Land Registry. Before putting the property up for sale, the Broker verifies the Seller's ownership certificate. The Broker offers the property in the form of a credential and establishes a price that the potential Buyer will have to pay. The Seller's Bank exchanges the mortgage deeds for cash with the Buyer's bank. The Buyer's Bank tells to the Registry to indicate the new possession by presenting the mortgage deeds thus, terminating the process.

Fig. 3. Chromaway case study choreography diagram.

The tool was implemented as a web application based on JavaScript programming language through the use of Node-Js and connected to the von-network which is a pre-packaged blockchain network of Hyperledger Indy[2].

Infrastructure Initialisation. In the initialisation phase, the developer prepares the necessary configuration to ensure a successful launch of the software components. In particular, this phase involves identifying and configuring agents that represent the actors of the choreography. This is done by specifying some information such as endpoints and DIDs in the related files inside the ChorSSI project. These agents could be human actors or automated systems, such as bots or APIs. After the manual set-up of the technical infrastructure, the ChorSSI built-in functionalities execute the successive low-level steps:

- **network setup**: builds the Docker container of the Indy network and starts the related nodes.
- **tail server setup**: Indy Tails Server is a specialised file server used to receive, store, and distribute Hyperledger Indy Tails files. Those contain the information required for credential holders to produce a zero-knowledge proof when responding to a proof request;
- **agents setup**: this step starts the agents of all the actors that participate in a choreography. Each participant is assigned to a specific agent with a specific set of parameters, according to the role of the participant.

Once all the mentioned steps have been performed, the script starts the ChorSSI web application. This application controls the various agents enabling

---

[2] https://github.com/bcgov/von-network

their communication with the blockchain and the participants involved in the system. Notice, the configured artefacts represent the technical infrastructure later bonded to choreography elements.

Modelling. In the modelling phase, the end-user can design the Choreography diagram used for executing SSI-related operations. The model contains all those interactions related to SSI, including the issuance of an identity and the verification of a credential. The modeller is contained in the ChorSSI home page and is based on the chor-js component [15] supporting the BPMN choreography elements. This makes it possible to create choreography diagrams in a standard manner, by using drag-and-drop functionalities to insert the desired elements. When modelling a choreography in ChorSSI, some aspects should be considered. First of all, message names have to correspond to the protocol names specified in the configuration. This facilitates recognising the SSI operation to be performed during the execution phase. Similarly, also choreography participants must be defined according to the agents' names in the initial configuration. The last newly included aspect consists of the extension of the modeller by including a new panel for the selection and insertion of SSI-related data. This panel is accessible by selecting a message element of the model. This interface allows the dynamic selection of SSI-related data objects to be exchanged during each phase of the execution, as well as the ability to define their content.

Participants Initialisation. After the choreography model is created, the initialisation of the participants produces the code necessary to connect the different parties and create the first artefact needed for the execution. In ChorSSI, this step is reported in Fig. 4 and it is executed by clicking on the start event of the choreography. Here the extended property panel includes an "SSI" tab, where it is possible to manage the participant initialisation. This step connects the agents defined in the initial configuration with the choreography participants, representing the organisations. Furthermore, the connection channels among them are automatically created. In particular, during this phase two main functionalities are available. The **connect participant** connects all the participants with

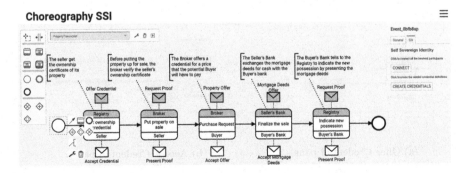

Fig. 4. Click on the model start event.

the underlying agents as indicated in the infrastructure configuration exploiting the dedicated API. The **create credential** instead creates the credential definition needed to execute all the tasks of the choreography starting from the "schema" previously defined.

Execution. In the execution phase, all the agents are connected among them with the relative credential definitions. After that, the choreography workflow can start with the end-user executing the first task of the choreography. In Fig. 5 is represented the *issuing* protocol of the Chromaway example which is composed of two main actions. Selecting the *Offer credential* message, the Registry participant acts as the issuer of the credentials. In the panel in Fig. 5a, the Registry can build and send the credential object. The first drop-down element selects the connection identifier of the Seller to issue the credential while the second one selects the credential definition previously created. This is used to create a credential object reported in the text area that can be customised. Once the needed information is set, the Registry can send the credentials using the related button. Here, ChorSSI automatically invokes the agent interacting with the blockchain. At this point, the Seller receives the credentials, becoming the holder that has to accept it. To do this, the Seller has only to click on the *Accept Credential* message of the same task. In this case, the panel in Fig. 5b only contains a drop-down element for the view of the received credentials. If the issued credential suits, it can be accepted by clicking the button that triggers the Agent API linked to the message with the parameters chosen from the panel. At this point, the issued credential is automatically stored in the Holder wallet and the sent message becomes green coloured indicating the completion of the task. The second task of the choreography diagram represents the *request proof* protocol with the initial execution made by the Broker participant. During this phase, the Broker sends the *Request Proof* message to the Prover role that is covered by the Seller participant. The panel contains a first selection of the Prover to whom send the request proof. The modifiable text area contains the default request proof object that can be modified. This object indicates in the

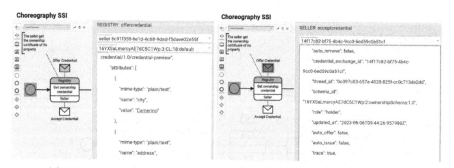

(a) Offer Credential panel.   (b) Accept Credential panel.

Fig. 5. Execution of *offer credential* protocol.

requested_attributes field all the attributes the Verifier wants to verify. Once the Request Proof is sent the corresponding Agent is invoked. The Seller executes the next step in the *Accept Credential* message. Initially, Prover has to respond with a presentation proof and then the Prover selects a request proof from the wallet. When the credential to present is filled, it can be sent to Broker. This triggers the Agent linked to the message with the parameters we have chosen from the panel. This concludes the execution of the second choreography task.

The last admitted SSI-related operation is the *revoke* of an issued credential. This can be done by an Issuer (e.g., Registry) clicking on the button on the page used also for monitoring purposes. In the case of the Holder (e.g., Seller), the credential previously owned becomes marked as *Revoked*. This can still be used for presenting proof but the Verifier will notice that the presented credential has been revoked.

For the sake of space, we reported here only the first two tasks of the diagram beyond all the activities included in the Chromaway example, but they represent all the operations admitted in an SSI scenario. Indeed, all the subsequent tasks in the choreography consist of the issuing of a credential or requesting proof like the already described tasks. The only possible change is the content of the default offered credential. For instance, in the case of the *Purchase Request* task, the credential definition is different and consequentially the set of attributes involved in the credential change.

Monitoring. The last phase of the ChorSSI methodology is the monitoring of the various operations performed during the choreography execution. For checking the content of the responses each user can navigate to the "Profile" section and select the related page. On the Profile page, we can find a status bar containing all the agents. From this, we can select the agent correlated to the participant of the diagram of which we want to observe information details (coloured in green).

The shown information will depend on the object that the selected agent has generated and the role played in the SSI system. If the selected agent is an issuer, the created credential definition, the linked schema object, and a reference to the issued credential are visible. Moreover, the possibility of revoking a credential appears. If the agent is a Holder the issued credential stored in its personal wallet is visible. If the agent is a verifier the objects regarding the verification of the proof presented by the holders are visible. Note that an agent could act as both a verifier and an issuer and a holder without constraints, it depends on the specific use case model.

An example of the monitoring page is reported in Fig. 6 which shows four objects since the Registry acts both as Issuer and Verifier in the proposed case study. The objects related to the role of Issuer played by the Registry agent are three: the **created schema** card showing the id, the name, and all the attributes contained in the schema, the **credential definition** created starting from the schema and finally the **issued credential**. This also references the credential that the agent issued and with a button gives the possibility to revoke the aforementioned credential.

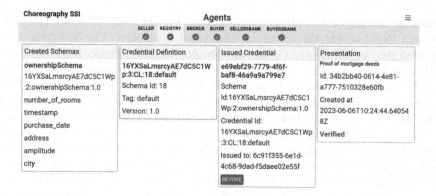

Fig. 6. View for Registry participant.

The seller agent contains instead the credential taken from the wallet that has been issued by the Registry. The credential contains the associated credential definition and schema IDs, the set of attributes with the associated values, and finally, a text showing that the credential has not been revoked. Also, the seller contains the presentation object showing the title of the requested proof, the unique identifier of the presentation, the date of creation, and a text showing if the verification has successfully ended.

4.1 Discussion

Analysing the ChorSSI approach and the proposed prototype, several considerations can be made. Among the benefits, certainly the first consists of the possibility of executing SSI operations without the need to implement each interaction since ChorSSI already provides the required built-in functionalities. Here the only effort resides in the initial infrastructure configuration in which the objects to exchange have to be defined. After that, the use of a choreography abstracts from the technicalities and only the needed information has to be inserted by the user. Indeed, the choreography defines the execution flow without the need to implement it. This is possible thanks to ChorSSI, which binds the BPMN messages to the standard built-in SSI protocols. A further benefit to consider is the support for monitoring purposes in an easy manner, without having to rely on external instruments. Indeed, ChorSSI already retrieves automatically all the exchanged information, providing it as a view to the end-user. Despite the above-mentioned benefits, there are still some minor aspects that can be improved. The first aspect concerns the infrastructure configuration which requires a certain level of expertise since it is still not fully integrated into the model. In the future, we intend to incorporate this aspect into the model so as to provide complete support in developing SSI systems. The last aspect concerns the insertion of information during the execution phase. Indeed, the data object panel presents the information in a JSON format. This can affect the user experience of the end-users. To overcome this limitation in the future we consider improving the user interface for data objects. Notice, in ChorSSI, we mainly focused on the functional aspects and the presented limitations do not affect the objective of the conceptual proposal. Anyway in the future we plan to cope with the presented limitations, providing an improved version of ChorSSI.

5   Related Works

In recent years, the field of self-sovereign identity has gained significant attention as it aims at giving individuals more control over their data and identities by allowing them to create and manage their own digital identities without relying on centralised authorities. In particular, this was supported by the use of blockchain which offers several advantages to SSI approaches [8,16]. In this context, the creation of supporting mechanisms for the development and execution of such systems is a relevant topic. This can be achieved through the use of visual models, which allow individuals with domain knowledge, but limited technical capabilities, to understand the content of each layer of the SSI ecosystem. This is beneficial as it allows for a wider range of stakeholders to participate in the development process and contribute their expertise [17].

However, while many works faced the development of blockchain-based systems using high-level models [18–21] exploiting model-driven techniques, the integration in the SSI domain is a relatively new and evolving field of research and few approaches are currently available. In [22], the work proposes a decentralised approach to private data storage and sharing, called the Private Data System (PDS), to address privacy issues arising from the centralised architectures used by governments and large companies to store and share massive amounts of personal data. PDS is composed of nodes spread across the Internet that communicate through executable choreographies, enabling self-sovereign storage and sharing of private data. While in PDS choreographies are used for achieving the communication between network nodes, in ChorSSI they are used at a higher level of abstraction, to act as a standard interface for end-users. Differently from ChorSSI, PDS is intended to address the self-sovereignty elements that arise from privacy regulations, rules, and principles. In [23] the authors present a platform that uses swarm communication and executable choreographies to implement smart contracts and a generic architecture for blockchain-based systems. The authors propose novel concepts, including secret smart contracts and the near-chain approach, a storage strategy to achieve data self-sovereignty. In this case, the proposed approach exploits a model-driven technique to provide a trusted environment for the creation and execution of smart contracts. In addition, the concept of self-sovereignty is persecuted with the use of Cloud Safe Boxes (CSB), an encrypted folder that can be seen as a sort of offline blockchain with special access rules about who can update or even delete them. With the introduction of choreographic smart contracts, CSB is exploring transaction validation without running the code in all nodes. Despite the required privacy of an SSI system is guaranteed and the performance is improved, the main difference with ChorSSI is the use of the choreography diagram to separate the technical complexity of an SSI system and to make it usable to non-technical organisations.

Although it is known that the use of visual approaches enables the easier development of SSI systems, concrete proposals supporting the modelling and execution through models are still lacking. For these reasons, ChorSSI introduces novelty by enabling the execution of SSI systems through choreographies, even for non-technical organisations.

6 Conclusions and Future Works

The continuous interconnection between people, things and services has raised the need for systems capable to store and manage digital identities. To this aim, over the years several identity models have been proposed, though they are usually centralised and rely on trusted third parties. A prominent alternative to those models is Self-sovereign Identity in which individuals are responsible for their own data. This has been enabled by blockchain, which provides a secure and decentralised infrastructure for implementing SSI systems. However, the development and execution of such software require technical knowledge and skills in different technologies from both developers and stakeholders, hindering the adoption of the SSI model.

For this reason, in this work, we propose ChorSSI, a BPMN-based framework for supporting the development and guiding the execution of Self-Sovereign Identity systems and functionalities. The proposed framework enables individuals and organisations to have control over their digital identity and to selectively disclose verifiable claims to other parties, without the need for intermediaries.

The use of BPMN choreography diagrams representing SSI protocols allows reducing the effort required for implementation and guides the correct execution flow among non-expert users. In particular, ChorSSI permits users to create and manage digital identities, issue and verify credentials, and share verifiable claims with other parties. The framework was evaluated by developing a prototype and executing the real-world Chromaway property transaction case study.

In future work, we intend to enrich the ChorSSI modelling phase by including additional details on the configuration so as to encapsulate the initialisation phase in a single initial choreography. Another point refers to the implementation of multi-tenancy agents, which makes it possible to represent multiple organisations for a single role. This makes it possible to support a wider range of BPMN elements such as, for instance, multiple instance ones.

References

1. Baars, D.: Towards self-sovereign identity using blockchain technology (2016)
2. Van Bokkem, D., Hageman, R., Koning, G., Nguyen, L., Zarin, N.: Self-sovereign identity solutions: The necessity of blockchain technology. arXiv preprint arXiv:1904.12816 (2019)
3. Tobin, A., Reed, D.: The inevitable rise of self-sovereign identity. Sovrin Found. **29**(2016), 18 (2016)

4. Bandara, E., Liang, X., Foytik, P., Shetty, S., De Zoysa, K.: A blockchain and self-sovereign identity empowered digital identity platform. In: 2021 International Conference on Computer Communications and Networks, pp. 1–7. IEEE (2021)
5. Grüner, A., Mühle, A., Gayvoronskaya, T., Meinel, C.: A comparative analysis of trust requirements in decentralized identity management. In: Barolli, L., Takizawa, M., Xhafa, F., Enokido, T. (eds.) AINA 2019. AISC, vol. 926, pp. 200–213. Springer, Cham (2020). https://doi.org/10.1007/978-3-030-15032-7_18
6. Belchior, R., Putz, B., Pernul, G., Correia, M., Vasconcelos, A., Guerreiro, S.: SSIBAC: self-sovereign identity based access control. In: International Conference on Trust, Security and Privacy in Computing and Communications, pp. 1935–1943. IEEE (2020)
7. Der, U., Jähnichen, S., Sürmeli, J.: Self-sovereign identity – opportunities and challenges for the digital revolution. arXiv preprint arXiv:1712.01767 (2017)
8. Ferdous, M.S., Chowdhury, F., Alassafi, M.O.: In search of self-sovereign identity leveraging blockchain technology. IEEE Access 7, 103059–103079 (2019)
9. Sedlmeir, J., Smethurst, R., Rieger, A., Fridgen, G.: Digital identities and verifiable credentials. Bus. Inf. Syst. Eng. 63(5), 603–613 (2021)
10. Kulabukhova, N.: Zero-knowledge proof in self-sovereign identity. In: CEUR Workshop Proceedings, vol. 2507, pp. 381–385 (2019)
11. Manski, S.: Distributed ledger technologies, value accounting, and the self sovereign identity. Front. Blockchain 3, 29 (2020)
12. OMG: Business process model and notation (BPMN) (2011). https://www.omg.org/spec/BPMN/2.0/PDF/
13. Mühle, A., Grüner, A., Gayvoronskaya, T., Meinel, C.: A survey on essential components of a self-sovereign identity. Comput. Sci. Rev. 30, 80–86 (2018)
14. David, A., Maciej, S., Lorenzino, V., Francesco, P.: Blockchain for digital government. Scientific analysis or review, Anticipation and foresight KJ-NA-29677-EN-N (online), KJ-NA-29677-EN-C (print), KJ-NA-29677-EN-E (ePub), Luxembourg (Luxembourg) (2019)
15. Ladleif, J., von Weltzien, A., Weske, M.: CHOR-JS: a modeling framework for BPMN 2.0 choreography diagrams. In: ER Forum/Posters/Demos, pp. 113–117 (2019)
16. van Bokkem, D., Hageman, R., Koning, G., Nguyen, L., Zarin, N.: Self-sovereign identity solutions: the necessity of blockchain technology (2019)
17. Sroor, M., Hickman, N., Kolehmainen, T., Laatikainen, G., Abrahamsson, P.: How modeling helps in developing self-sovereign identity governance framework: an experience report. Procedia Comput. Sci. 204, 267–277 (2022). International Conference on Industry Sciences and Computer Science Innovation
18. Corradini, F., et al.: Model-driven engineering for multi-party business processes on multiple blockchains. Blockchain Res. Appl. 2(3), 100018 (2021)
19. Corradini, F., Marcelletti, A., Morichetta, A., Polini, A., Re, B., Tiezzi, F.: Engineering trustable and auditable choreography-based systems using blockchain. ACM Trans. Manag. Inf. Syst. 13(3), 31:1-31:53 (2022)
20. Corradini, F., Marcelletti, A., Morichetta, A., Polini, A., Re, B., Tiezzi, F.: A flexible approach to multi-party business process execution on blockchain. Futur. Gener. Comput. Syst. 147, 219–134 (2023)
21. López-Pintado, O., García-Bañuelos, L., Dumas, M., Weber, I., Ponomarev, A.: Caterpillar: a business process execution engine on the Ethereum blockchain. Softw. Pract. Exp. 49(7), 1162–1193 (2019)

22. Alboaie, S., Cosovan, D.: Private data system enabling self-sovereign storage managed by executable choreographies. In: Chen, L.Y., Reiser, H.P. (eds.) DAIS 2017. LNCS, vol. 10320, pp. 83–98. Springer, Cham (2017). https://doi.org/10.1007/978-3-319-59665-5_6
23. Alboaie, S., Alboaie, L., Pritzker, Z., Iftene, A.: Secret smart contracts in hierarchical blockchains (2019)

Loose Collaborations on the Blockchain: Survey and Challenges

Tom Lichtenstein[1]([✉]), Hassan Atwi[2], Mathias Weske[1], and Cesare Pautasso[2]

[1] Hasso Plattner Institute, University of Potsdam, Potsdam, Germany
{Tom.Lichtenstein,Mathias.Weske}@hpi.de
[2] Università della Svizzera Italiana, Lugano, Switzerland
{Hassan.Atwi,Cesare.Pautasso}@usi.ch

Abstract. Blockchain technology has emerged as a promising infrastructure for enabling collaboration between mutually distrustful organizations. The enactment of blockchain-based collaborative processes typically requires a profound understanding of the process being executed, limiting support for flexible processes that cannot be fully prespecified at design time. To overcome this limitation, support for looseness, dealing with the configuration and execution of underspecified processes, is essential. In this paper, we conduct a systematic literature review to examine looseness support for blockchain-based collaborative processes from a behavioral and organizational perspective. In addition, we identify open research challenges to pave the way for further research in this area.

Keywords: Collaborative processes · Flexibility · Looseness · Blockchain

1 Introduction

The adoption of blockchain technology in business process management (BPM) opened up new opportunities for the collaborative execution of business processes. Unlike traditional, centralized BPM systems for process orchestration, blockchain technology provides a trusted environment for the decentralized execution of collaborative processes [20]. Most current blockchain-based approaches for process execution require a comprehensive understanding of the process prior to deployment, typically using a model-driven design approach [26]. However, in knowledge-intensive domains, processes are required to support dynamic behavior that cannot be fully anticipated at design time, thus demanding a more flexible execution environment [5]. To support the execution of such dynamic processes on blockchains, support for looseness is required.

The concept of looseness deals with the configuration and execution of underspecified processes. By incorporating looseness support, process specifications can be refined beyond design time [24]. However, due to the immutability of the deployed logic, associated costs, and increased complexity, deferring refinements can pose challenges in a blockchain environment. This survey investigates the support of looseness for the execution of blockchain-based collaborative processes based on the following research questions:

© The Author(s), under exclusive license to Springer Nature Switzerland AG 2023
J. Köpke et al. (Eds.): BPM 2023, LNBIP 491, pp. 21–35, 2023.
https://doi.org/10.1007/978-3-031-43433-4_2

RQ1: To what extent do model-driven blockchain-based approaches for executing collaborative processes support looseness in existing literature?

RQ2: What are open challenges in supporting looseness in collaborative processes on the blockchain?

To answer the research questions, this paper presents a systematic literature review analyzing looseness support of current approaches for blockchain-based collaborative business process execution from a behavioral and organizational perspective. Based on the results, open research challenges are derived.

In the following, Sect. 2 outlines fundamental concepts such as collaborative process execution using blockchain technology and looseness, as well as related work. Next, Sect. 3 presents the methodology used for the systematic literature review. Section 4 reviews the current looseness support of approaches enabling blockchain-based collaborations based on the selected literature. Finally, Sect. 5 identifies open challenges, and Sect. 6 concludes this paper.

2   Foundation and Related Work

2.1   Blockchain-Based Collaborative Processes

In an interorganizational environment, organizations are entailed to collaborate by bridging their internal workflows to create value that equitably benefits all participants. The collaboration can take different forms according to the desired goal and boundary scope for each organization [19]. Business process management offers a set of methods to design, execute, and optimize business processes to reduce the ambiguity of complex cross-organizational interaction behavior [32]. The literature refers to interacting processes involving multiple participants with different terms, such as process collaboration or choreography. In the context of this paper, we adopt the term *collaborative process* to encompass processes of interorganizational nature [6].

Process-Aware Information Systems (PAIS) ensure that the participants adhere to the execution order of activities by keeping track of the execution state of a process. Centralized PAIS architectures are intended to support the orchestration of internal processes within the boundaries of an organization [21]. However, collaborative processes are decentralized in the sense that they are not controlled by a single authority. To overcome this challenge, traditional PAIS architectures must evolve to support collaborative business processes.

The emergence of blockchain technology, enabling the creation of a secure and distributed ledger among a network of participants without the need for mutual trust, holds great potential as a platform for executing and monitoring collaborative business processes [6,30]. Smart contracts emerged as a viable approach to execute business logic on the blockchain [31]. The development of blockchain-based collaborative processes typically follows a model-driven approach, translating a process model into an executable code embedded in smart contracts. Subsequently, collaborating participants engage with the smart contracts by initiating transactions, thereby advancing the state of the process.

2.2   Looseness in Business Processes

In knowledge-intensive domains, it is often not possible to fully specify entire business processes in advance due to their unpredictable, non-repeatable and emergent nature [5]. For such situational processes, only certain parts are known a priori. Consequently, the process models used to represent them are typically underspecified. The execution of these processes on a PAIS requires support for *looseness* [2]. Along with *variability, adaptation,* and *evolution,* looseness is one of the four flexibility requirements of PAISs that deals with supporting the configuration and execution of underspecified processes [24]. Unlike adaptation and evolution, where changes are made to the prespecified parts of a process, looseness focuses on refining parts that lack specifications in the original model. The need for looseness can be observed from different *process perspectives.* This study focuses specifically on the *behavioral* and *organizational* perspective, which consider execution behavior, i.e., control flow, and the assignment of actors to tasks, respectively. Other perspectives, including operational, functional, informational and temporal aspects, are beyond the scope of this paper.

Achieving looseness in collaborative processes is particularly challenging, since compatibility between all participants must be maintained during refinement. Blockchain technology can address this challenge by providing a decentralized infrastructure that tracks and enforces refinement decisions. Figure 1 illustrates different refinement patterns for loosely specified processes along the lifecycle phases of blockchain-based collaborative processes. The lifecycle is derived from [12] and extended with observations from the literature. In general, we expect the lifecycle to begin with the creation of a model that is compiled into a blockchain-readable format that can then be deployed. Afterward, the process can be instantiated and executed.

According to Fig. 1, processes that do not support looseness follow the *fully prespecified* pattern. This pattern implies that all information needed for process execution is already specified in the modeling phase, and no refinements are required afterward. In contrast, if the model still contains underspecified parts

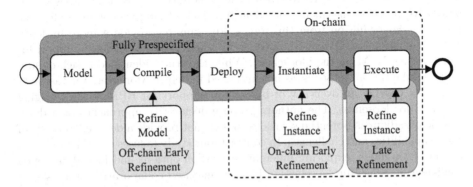

Fig. 1. Looseness patterns applied to the corresponding phases of an extended lifecycle of blockchain-based collaborative process inspired by [12].

after the modeling phase, the *early* or *late refinement* pattern can be applied. The early refinement pattern allows the process to be refined in between the initial modeling phase and the execution phase. Here, the process can be configured according to the needs of a specific instance to be created. In a blockchain setting, we distinguish between *off-chain early refinement*, where additional information is fed to the off-chain compiler to produce the blockchain artifacts required for execution, and *on-chain early refinement*, where the information is provided on-chain during instantiation. Finally, *late refinement* allows for the gradual refinement of collaborative processes during the execution phase.

For a more detailed classification of looseness support, we adopt the *decision deferral taxonomy* proposed by Reichert and Weber [24]. This taxonomy includes, in addition to the process perspective, five dimensions to assess the level of looseness supported by a PAIS: (i) *degree of freedom*, (ii) *planning approach*, (iii) *scope of deferral*, (iv) *degree of automation*, and (v) *decision-making*.

In this context, the degree of freedom refers to the flexibility in choosing the refinement of underspecified parts. The planning approach indicates the degree of prespecification required and the life cycle phase to which refinement decisions can be deferred. The scope of deferral describes the extent to which a process is affected by the need for refinement. The degree of automation reflects the support provided by the underlying information system to automate the refinement of loose specifications. Decision-making determines the primary indicator used to decide on refinements. Table 1 presents the characteristics of each dimension.

2.3 Related Work

Existing literature has examined the relevance of process flexibility and its impact on BPM. Cognini et al. [2] present a comprehensive study that discussed the impact of flexibility on different phases of the BPM lifecycle. However, their study does not consider blockchain technology. Mendling et al. [20] focus on the opportunities and challenges of blockchain-based process execution, emphasizing the need for adaptation and evolution. Looseness is not specifically addressed. In [26], Stiehle and Weber investigate the capabilities of blockchain-based collaborative process enactment and identify a lack of support for unpredictable processes. Following on from this, our study aims to provide a more in-depth analysis of looseness. Garcia-Garcia et al. [6] assess the BPM lifecycle support for collaborative BPM on the blockchain. While the study identified a lack of support for adaptation, looseness is not discussed in depth. Moreover, Viriyasitavat et al. [30] provide an overview of blockchain support for BPM, acknowledging the need for looseness. The authors propose declarative approaches or a data-centric paradigm as potential solutions, but do not provide a detailed analysis of their capabilities and challenges. In contrast, our study focuses on addressing the aspect of looseness, exploring different realization options and identifying open challenges in achieving looseness in blockchain-based collaborative processes.

Table 1. Looseness dimension characteristics derived from [24].

Dimension	Character.	Description
Degree of freedom	None	All aspects are prespecified by the model (fully prespecified pattern).
	Selection	The model includes underspecified parts and provides a predefined set of options to complete the business logic of the process.
	Modeling & Composition	The process is loosely composed from known components following predefined constraints.
Planning approach	Plan-driven	After modeling phase, the underspecified parts must be refined before entering the execution phase (early refinement pattern).
	Iterative	After modeling phase, the unspecified parts can be refined during the execution phase (late refinement pattern).
	Ad-hoc	No modeling phase takes place in advance. The process is designed during execution phase.
Scope of deferral	Regional	The process contains prespecified parts. Only certain regions are loosely-specified and require refinements.
	Entirety	The process follows no predefined schema. The need for refinements extends throughout the entire process.
Degree of automation	Manual	Refinements require user intervention. The system does not support the user in deciding on refinement options.
	System-supported	Refinements require user intervention. The system provides the user with information or functionality that supports the refinement process.
	Automated	Refinement decisions are made automatically. No user intervention is required.
Decision-making	Goal-based	Refinement decisions must ensure that predefined goals can be achieved.
	Rule-based	Refinement decisions are constrained by predefined rules.
	Experience-based	Refinement decisions are based on the results of previous executions related to the current execution.
	User-based	Refinement decisions depend only on the end user.

3   Systematic Literature Review Methodology

To address the research questions outlined in Sect. 1, a systematic literature review is conducted. Relevant literature is identified based on three groups of keywords, forming the following search query:

```
("Blockchain" OR "Distributed Ledger" OR "Smart Contract") AND
("Business Process" OR "Workflow" OR "Choreography" OR "BPM") AND
("Flexible" OR "Loose" OR "Dynamic" OR "Declarative")
```

Table 2. Inclusion and exclusion criteria used for the systematic literature review.

Inclusion Criteria	Exclusion Criteria
1. The study presents or extends a model-driven approach enabling the blockchain-based enactment of collaborative processes. 2. The study discusses concepts linked to behavioral or organizational looseness.	1. The study uses blockchain technology for monitoring purposes only. 2. The study is not primary research, e.g., a literature review, survey, or overview paper. 3. The study does not qualify as a research paper, e.g., patents, technical specifications. 4. The study is not written in English.

Based on the search query, studies are collected from reputable academic databases, including *IEEE Xplore, ScienceDirect, SpringerLink, Wiley Online Library*, and *Google Scholar*. Due to the contemporary nature of the blockchain area, gray literature is considered as well. Initially, the search yielded a total of 2147 studies on 05/09/2023. In a three-step process, the relevance of these studies is assessed based on their title, abstract, and full text. The criteria used to determine their inclusion or exclusion are described in Table 2. In case of the inclusion criteria, it should be noted that studies must not explicitly mention the support of looseness, but may describe concepts that can be mapped to looseness in the corresponding perspectives. In contrast, studies that focused only on process monitoring were excluded, as we focus on execution support for blockchain-based collaborative processes.

After applying the criteria, a total of 24 papers are identified as relevant to the study. They include eight journal articles, 13 conference papers, one workshop paper, one symposium paper, and one gray literature paper, which serve as the basis for the analysis of the research questions in the following sections.

4 Looseness in Collaborations on the Blockchain

In this section, we classify the studies under review based on the decision deferral taxonomy described in Table 1, focusing on the behavioral and organizational process perspectives.

Figure 2 depicts the distribution of the selected literature in terms of its support for looseness in the two process perspectives. The distribution indicates a focus on support for organizational looseness. Furthermore, Fig. 3 shows the looseness patterns applied by the studies for each perspective. In accordance with the descriptions in Sect. 2.2, we examine the characteristics of each study regarding each dimension. The results of this review are presented in Table 3, which provides a comprehensive categorization of the studies. It should be noted that studies with a degree of freedom of 'None' are not further assessed for other dimensions within the same process perspective, as they do not indicate support for looseness from the corresponding perspective. Due to the focus on model-driven approaches, Fig. 4 provides an overview of the modeling languages used by the studies. Consequently, approaches supporting ad-hoc planning are outside the scope of this survey, as they omit any upfront modelling. In the following, we provide a detailed analysis of the results from each process perspective to gain insights into the current support for looseness in model-driven collaborative processes using blockchain to answer RQ1.

4.1 Behavioral Perspective

When examining looseness from a behavioral perspective, we investigate how the selected approaches enact a loosely defined control-flow on the blockchain.

Degree of freedom deals with the flexibility in assigning concrete business logic to unspecified parts, e.g., placeholders. A collaborative process with a fully prespecified activity sequence is generally considered not to retain any aspects of behavioral looseness. The conducted literature review has shown that 14 studies do not support any *degree of freedom* in their process. The aforementioned studies define their collaborative process during the modelling phase and leave no latitude to a loose control-flow. While the studies do not support looseness from a behavioral perspective, they can still support looseness in other process perspectives, such as the organizational perspective (16 studies). Eight studies do implement some *degree of freedom* in their process. Of these, three studies support process refinement through selection, while the remaining five do so through modeling & composition. In the selection strategy, a set of process fragments are predefined during the modelling phase. The fragments are either stored in a repository off-chain [23] or deployed on-chain as individual smart

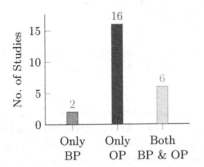 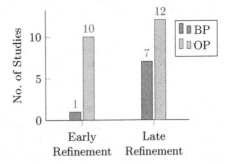

Fig. 2. Looseness support of studies based on process perspectives. Studies can support looseness from only an organizational (OP) or behavioral perspective (BP), or support both.

Fig. 3. Refinement patterns employed by studies according to the behavioral (BP) and organizational perspective (OP).

Table 3. Looseness support provided by the selected literature from a behavioral and organizational perspective.

Decision Deferral		Behavioral Perspective		Organizational Perspective	
Dimension	Character.	No.	Reference List	No.	Reference List
Degree of Freedom	None	16	[3,4,8–11,13,14,16,17,22,27–29,31,35]	2	[33,34]
	Selection	3	[15,23,34]	21	[1,3,4,7–11,13–18,23,25,27–29,31,35]
	Mod. & Com.	5	[1,7,18,25,33]	1	[22]
Planning Approach	Plan-driven	1	[23]	10	[1,7,10,11,17,18,23,27,28,31]
	Iterative	7	[1,7,15,18,25,33,34]	12	[3,4,8,9,13–16,22,25,29,35]
	Ad-hoc	0		0	
Scope of Deferral	Regional	3	[15,23,34]	2	[8,29]
	Entirety	5	[1,7,18,25,33]	20	[3,4,7,9–11,13–18,22,23,25,27,28,31,35]
Degree of Automation	Manual	0		14	[1,3,4,7,11,16–18,22,25,27,28,31,35]
	Sys.-sup.	7	[1,7,15,18,25,33,34]	5	[9,10,13–15]
	Automated	1	[23]	3	[8,23,29]
Decision Making	Goal-based	0		1	[23]
	Rule-based	7	[1,7,15,18,23,25,33]	3	[13–15]
	Exp.-based	0		2	[8,29]
	User-based	1	[34]	16	[1,3,4,7,9–11,16–18,22,25,27,28,31,35]

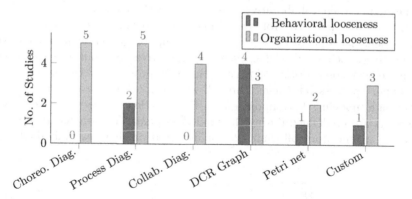

Fig. 4. Modeling languages used by studies related to looseness support based on the perspectives. A study may provide looseness support for both perspectives.

contracts [15,34] ready to be reused during the process refinement. On the other hand, studies adopting modelling & composition follow a fine-grained placeholder refinement by composing fragments using individual activities. Imperative modelling languages like BPMN are not as expressive for such a purpose as declarative approaches. For instance, we find that four studies supporting modelling & composition make use of *dynamic condition response* (DCR) graphs to express the control-flow [1,7,18,33]. Furthermore, [25] adopts *artifact-centric processes*, which provide a declarative approach for specifying the control-flow.

In the *planning approach*, we examine the phases in which the control-flow can be refined in the lifecycle of blockchain-based collaborative processes. In this context, we focus on the phases depicted in Fig. 1. Following a plan-driven strategy, the refinement of the control-flow takes place prior to the execution phase, i.e., early refinement pattern. According to the results, one study–a framework for modelling smart contract control-flow [23]–is compliant with this strategy. The proposed framework makes use of a set of predefined fragments represented as individual smart contracts to generate the workflow of a collaborative process. Since the process of generating the workflow takes place in an off-chain phase, this could be considered as an off-chain early refinement according to the lifecycle in Fig. 1. On the other hand, seven studies adopt an iterative strategy, in which the refinement of a control-flow happens while the process is being executed on-chain. For instance, in [7], the author embeds the DCR execution rules into a smart contract, ensuring the correctness of process refinement at run-time, an example of the late-refinement pattern.

Scope of deferral specifies the loosely defined region of a control-flow. Five studies allow looseness in the entirety of the process [1,7,18,25,33]. These studies rely on declarative process representations, such as DCR graphs and artifact-centric processes. The remaining three studies limit behavioral looseness to a specific region of the control-flow. For instance, in [15,23,34] the main process is prespecified, while certain regions are left to be refined later.

Placeholder assignment can be done through various *degrees of automation*. No approach adopts a manual degree of automation. Since manual refinements can lead to compatibility issues, refinement of collaborative processes can benefit from system support. In the literature, seven studies rely on system support for the refinement of processes. For example, [34] addresses the issue of process state inconsistency that may be a consequence of a control-flow refinement event. The authors employ fragments as individual smart contracts under the name of navigators which are attached to the process instance at run-time, the navigator must fulfill some preconditions before a successful attachment. Only one study [23] adopts fully automated refinement by generating a workflow composed of multiple smart contracts without the intervening of participants.

Upon placeholder activity enablement, numerous decisions have to be taken in order to proceed with the process execution. *Decision-making* can be based on rules governing how the control-flow is allowed to behave. Seven of the selected studies support rule-based decision-making. Most notably, [15] introduces agreement policies, which restrict how participants can refine control-flow elements. Furthermore, the request to refine the control-flow has to be endorsed by participants during runtime through a voting process embedded in a smart contract. The *decision-making* can also be taken *by participants*. For instance, in [34], a specific participant, namely the process coordinator, is responsible for control-flow changes. While [34] provides rules for process refinement, they only assure consistency rather than restricting the reasoning behind a refinement. Goal-based *decision-making* is not present in any of the studies, as none of the approaches rely on goal specifications to achieve control-flow refinement.

In response to RQ1, the study found that the concept of looseness, particularly in terms of behavioral looseness, has received limited attention in existing literature. For instance, no wide support for automated process refinement could be observed. In addition, the literature lacks a discussion of experience-based approaches, where the transaction log is used to make optimal control flow refinement decisions, and goal-based approaches.

4.2   Organizational Perspective

Organizational looseness refers to the refinement options available for managing resources after the modeling phase. In collaborative processes, this primarily concerns the assignment of participants to tasks [24]. In the following, we distinguish between the terms *actor* and *role*. An actor represents an actual entity involved in a collaboration, e.g., a specific organization, while a role serves as a descriptive placeholder that can be associated with one or more actors.

The *degree of freedom* determines the flexibility in assigning actors to tasks. A degree of freedom of none implies that the assignment of actors to tasks is predetermined by the model. Although no predetermined assignments were observed in the literature reviewed, two studies lacked sufficient insight into resource management to determine the presence of organizational looseness [33,34]. A degree of freedom of selection can be achieved by linking tasks to roles in the model instead of assigning actors directly. Thus, any actor associated with the corresponding role can be selected for the task. Since most of the process modelling languages used in the selected literature inherently include roles, such as BPMN collaboration [9,10,27] or choreography diagrams [3,4,11,16,31] (Fig. 4), selection is supported by the majority of studies. Modelling & composition can be achieved by omitting roles and allowing to select an actor freely for each task from a pool of known actors. In the literature reviewed, one study supports modelling and composition from an organizational perspective [22]: Actors that have completed a part of the choreography can transfer execution control to the next actor via a blockchain transaction. The selection of the next actor is left freely to the actor currently in control of the execution.

Regarding the *planning approach*, both plan-driven and iterative strategies are similarly represented in the selected literature. Plan-driven approaches bind actors to tasks prior to the execution phase, thus following an early refinement pattern. Two studies adopt an off-chain early refinement pattern [1,23], while eight studies use on-chain early refinement [7,10,11,17,18,27,28,31]. In contrast, iterative strategies allow late refinement by binding actors to tasks during the execution phase. While five iterative approaches allow late binding of actors only once [3,4,8,29,35], eight approaches also support rebinding actors during execution [9,10,13–16,22,25]. In addition, two studies allow specifying mandatory and optional roles [3,4], where the former must be refined at the instantiation phase and the latter can be selected during the execution phase.

Considering the *scope of deferral*, most approaches allow a loose binding for the entire process, implying that no actor is prespecified. However, two studies indicate a regional approach by focusing on binding only specific service providers [8,29]. While two studies distinguish between mandatory and optional roles [3,4], since both types of roles can be loosely specified using either early or late refinement, looseness is still provided for the entire process.

The *degree of automation* shows a preference for the manual binding of actors in the selected literature. However, five studies provide system support for refinement decisions. This includes voting smart contracts, which allow to collectively decide on the binding of actors on-chain [9,10]. Three studies use smart contracts to enforce policies that specify the conditions to be considered when binding actors [13–15]. In addition, one study offers a fully automated selection of actors required to execute an order [23], and two studies allow the automatic selection of service providers based on predefined criteria [8,29].

In terms of *decision-making*, most studies provide end-users with full control over actor binding. Rule-based approaches are employed by three studies, using predefined policies to restrict actor bindings [13–15]. Experience-based decision-making relies on previous executions and is supported by two studies [8,29]: The approaches select a service provider based on quality of service ratings from past executions stored on the blockchain. The most appropriate service provider can be selected at runtime using predefined filtering and sorting criteria. Finally, one study demonstrates goal-based *decision-making*, where collaboration participants are selected based on an incoming order to form a supply chain [23].

Addressing RQ1, organizational looseness has received extensive support in the selected literature. Given its widespread acceptance as a fundamental modeling concept, 21 studies used roles to allow loose binding of actors. 12 studies support late refinement to enhance runtime flexibility. However, 14 studies rely on manual and user-driven decision-making, thus providing no decision support.

5   Research Challenges

During our analysis, we identified several challenges that need to be considered when supporting looseness in blockchain-based collaborative processes. In this section, we aim to address RQ2 by exploring the challenges and discussing how to overcome them.

Refinement Support. Determining the authority responsible for refining underspecified parts in collaborative processes is challenging, as each collaborator may pursue their own interests. Especially from an organizational perspective, current approaches often follow manual and user-based decision-making, leaving refinement decisions to individual users. However, this strategy does not necessarily reflect the collective interests or ensure optimal outcomes. Lopez-Pintado et al. [13–15] introduce on-chain policies to control the refinement as a promising first step to address this challenge. In addition, Viriyasitavat et al. [29] and Henry et al. [8] use on-chain information from previous executions to guide organizational refinement. However, experience-based and goal-based behavioral refinement is not extensively studied, making them topics worthy of investigation. While initial approaches employ system-supported decision-making, automated decision-making remains largely unexplored. Therefore, looseness support can be enhanced by further research on blockchain-based process refinement support.

Cost Optimization. Introducing looseness support for processes executed on public blockchains can lead to additional costs, for example, due to the need to verify declarative rules for each task execution. Furthermore, looseness can be exploited as an attack vector, as adversarial actors can add refinements that force other parties to bear higher transaction costs [18]. Hence, it is essential to carefully weigh the benefits of looseness against added costs. In particular, the development of hybrid approaches that combine loose and structured specifications, allowing looseness only in certain areas, needs further investigation. Furthermore, to enable fair collaboration, cost-sharing approaches need to be explored. As a first step in this direction, Klinger et al. [9,10] propose a fair cost distribution mechanism for the deployment of new instances in their approach.

Loose Modeling Support. Model-driven development shows potential for designing and implementing collaborative processes using blockchain technology, as evidenced by its widespread adoption [26]. Since the existing literature focuses on highly structured modeling languages (Fig. 4), there is a notable lack of research on loose modeling languages for developing blockchain-based collaborations. While initial studies use DCR graphs to achieve behavioral looseness [1,7,18,33], additional languages need to be explored to enable the selection of the most appropriate modeling language for a given business case. In addition, the integration of blockchain-specific concepts [11] into these languages can foster model-driven development of loose blockchain-based collaborative processes.

Loose Implementation Patterns. Implementing loose collaborative processes on the blockchain is challenging due to the increased complexity compared to fully prespecified processes. To gain a comprehensive overview of how to effectively leverage blockchain properties to achieve looseness, a detailed analysis of blockchain implementation patterns is essential. The findings can serve as the basis for designing a guideline that facilitates the development of loose collaborations on the blockchain, eventually leading to more reliable applications.

In summary, to answer RQ2, there are several challenges to supporting looseness in blockchain-based collaborative processes. These challenges include support for refinement decisions, optimizing the costs associated with looseness, and developing appropriate modelling concepts as well as implementation patterns. Overcoming these challenges is critical to successfully realizing the benefits of looseness in blockchain-based collaborative processes.

6 Conclusion

This paper explores the current support (RQ1) and challenges (RQ2) in enabling looseness for blockchain-based collaborative business process executions, considering both behavioral and organizational perspectives. To this end, a systematic literature review is conducted to classify model-driven approaches providing looseness support. The selected studies are analyzed using the decision deferral taxonomy [24], which allows the assessment of their level of looseness support. The results indicate that current approaches already largely support organizational looseness, but lack support for behavioral looseness. Based on the findings, challenges related to refinement support, cost optimization, modeling support, and implementation patterns are identified that require further investigation to advance looseness support for blockchain-based collaborative processes.

Since RQ1 is only examined from an organizational and behavioral perspective, future work can explore looseness support from additional process perspectives mentioned in Sect. 2.2. Moreover, investigating the relationships between different blockchain implementation patterns and support for looseness dimensions could contribute to a comprehensive understanding of looseness-enabling patterns for collaborative business processes using blockchain technology.

Acknowledgement. Funded by the Deutsche Forschungsgemeinschaft (DFG, German Research Foundation) – 450612067

References

1. Brahem, A., et al.: A trustworthy decentralized change propagation mechanism for declarative choreographies. In: Di Ciccio, C., Dijkman, R., del Río Ortega, A., Rinderle-Ma, S. (eds.) BPM 2022, LNCS, vol. 13420, pp. 418–435. Springer, Cham (2022). https://doi.org/10.1007/978-3-031-16103-2_27
2. Cognini, R., Corradini, F., Gnesi, S., Polini, A., Re, B.: Business process flexibility - a systematic literature review with a software systems perspective. Inf. Syst. Front. **20**(2), 343–371 (2018)
3. Corradini, F., et al.: Model-driven engineering for multi-party business processes on multiple blockchains. Blockchain Res. Appl. **2**(3), 100018 (2021)
4. Corradini, F., Marcelletti, A., Morichetta, A., Polini, A., Re, B., Tiezzi, F.: Engineering trustable and auditable choreography-based systems using blockchain. ACM Trans. Manage. Inf. Syst. **13**(3), 1–53 (2022)
5. Di Ciccio, C., Marrella, A., Russo, A.: Knowledge-intensive processes: characteristics, requirements and analysis of contemporary approaches. J. Data Semant. **4**(1), 29–57 (2015)
6. Garcia-Garcia, J.A., Sanchez-Gomez, N., Lizcano, D., Escalona, M.J., Wojdynski, T.: Using blockchain to improve collaborative business process management: systematic literature review. IEEE Access **8**, 142312–142336 (2020)
7. Henry, T., Brahem, A., Laga, N., Hatin, J., Gaaloul, W., Benatallah, B.: Trustworthy cross-organizational collaborations with hybrid on/off-chain declarative choreographies. In: Hacid, H., Kao, O., Mecella, M., Moha, N., Paik, H. (eds.) ICSOC 2021. LNCS, vol. 13121, pp. 81–96. Springer, Cham (2021). https://doi.org/10.1007/978-3-030-91431-8_6
8. Henry, T., Laga, N., Hatin, J., Beck, R., Gaaloul, W.: Hire me fairly: towards dynamic resource-binding with smart contracts. In: Proceeding of International Conference on Services Computing (SCC), pp. 407–412. IEEE (2021)
9. Klinger, P., Bodendorf, F.: Blockchain-based cross-organizational execution framework for dynamic integration of process collaborations. In: Proceeding of WI, pp. 893–908. GITO Verlag (2020)

10. Klinger, P., Nguyen, L., Bodendorf, F.: Upgradeability concept for collaborative blockchain-based business process execution framework. In: Chen, Z., Cui, L., Palanisamy, B., Zhang, L.-J. (eds.) ICBC 2020. LNCS, vol. 12404, pp. 127–141. Springer, Cham (2020). https://doi.org/10.1007/978-3-030-59638-5_9
11. Ladleif, J., Weske, M., Weber, I.: Modeling and enforcing blockchain-based choreographies. In: Hildebrandt, T., van Dongen, B.F., Röglinger, M., Mendling, J. (eds.) BPM 2019. LNCS, vol. 11675, pp. 69–85. Springer, Cham (2019). https://doi.org/10.1007/978-3-030-26619-6_7
12. Liu, Y., Lu, Q., Yu, G., Paik, H., Perera, H., Zhu, L.: A pattern language for blockchain governance. In: Proceedings of EuroPLop, pp. 28:1–28:16. ACM (2022)
13. López-Pintado, O., Dumas, M., García-Bañuelos, L., Weber, I.: Dynamic role binding in blockchain-based collaborative business processes. In: Giorgini, P., Weber, B. (eds.) CAiSE 2019. LNCS, vol. 11483, pp. 399–414. Springer, Cham (2019). https://doi.org/10.1007/978-3-030-21290-2_25
14. López-Pintado, O., Dumas, M., García-Bañuelos, L., Weber, I.: Interpreted execution of business process models on blockchain. In: Proceedings of International Enterprise Distributed Object Computing Conference (EDOC), pp. 206–215. IEEE (2019)
15. López-Pintado, O., Dumas, M., García-Bañuelos, L., Weber, I.: Controlled flexibility in blockchain-based collaborative business processes. Inf. Syst. **104**, 101622 (2022)
16. Loukil, F., Boukadi, K., Abed, M., Ghedira-Guegan, C.: Decentralized collaborative business process execution using blockchain. World Wide Web **24**(5), 1645–1663 (2021). https://doi.org/10.1007/s11280-021-00901-7
17. Lu, Q., et al.: Integrated model-driven engineering of blockchain applications for business processes and asset management. Softw. Pract. Exp. **51**(5), 1059–1079 (2021)
18. Madsen, M., Gaub, M., Kirkbro, M., Høgnason, T., Slaats, T., Debois, S.: Collaboration among adversaries: distributed workflow execution on a blockchain (2018). https://pure.itu.dk/en/publications/collaboration-among-adversaries-distributed-workflow-execution-on, Presented at Symposium on Foundations and Applications of Blockchain 2018
19. Mandell, M., Steelman, T.: Understanding what can be accomplished through interorganizational innovations the importance of typologies, context and management strategies. Publ. Manage. Rev. **5**(2), 197–224 (2003)
20. Mendling, J., et al.: Blockchains for business process management - challenges and opportunities. ACM Trans. Manag. Inf. Syst. **9**(1), 4:1-4:16 (2018)
21. Pourmirza, S., Peters, S., Dijkman, R., Grefen, P.: A systematic literature review on the architecture of business process management systems. Inf. Syst. **66**, 43–58 (2017)
22. Prybila, C., Schulte, S., Hochreiner, C., Weber, I.: Runtime verification for business processes utilizing the bitcoin blockchain. Future Gener. Comput. Syst. **107**, 816–831 (2020)
23. Rahman, M.S., Khalil, I., Bouras, A.: A framework for modelling blockchain based supply chain management system to ensure soundness of smart contract workflow. In: Proceedings of Hawaii International Conference on System Sciences (HICSS), pp. 1–10. ScholarSpace (2021). https://hdl.handle.net/10125/71295
24. Reichert, M., Weber, B.: Enabling Flexibility in Process-Aware Information Systems - Challenges, Methods, Technologies. Springer, Heidelberg (2012). https://doi.org/10.1007/978-3-642-30409-5

25. Amaral de Sousa, V., Burnay, C., Snoeck, M.: B-MERODE: a model-driven engineering and artifact-centric approach to generate blockchain-based information systems. In: Dustdar, S., Yu, E., Salinesi, C., Rieu, D., Pant, V. (eds.) CAiSE 2020. LNCS, vol. 12127, pp. 117–133. Springer, Cham (2020). https://doi.org/10.1007/978-3-030-49435-3_8

26. Stiehle, F., Weber, I.: Blockchain for business process enactment: a taxonomy and systematic literature review. In: Marrella, A., et al. (eds.) BPM 2022. LNBIP, vol. 459, pp. 5–20. Springer, Cham (2022). https://doi.org/10.1007/978-3-031-16168-1_1

27. Sturm, C., Scalanczi, J., Schönig, S., Jablonski, S.: A blockchain-based and resource-aware process execution engine. Future Gener. Comput. Syst. **100**, 19–34 (2019)

28. Sturm, C., Szalanczi, J., Schönig, S., Jablonski, S.: A lean architecture for blockchain based decentralized process execution. In: Daniel, F., Sheng, Q.Z., Motahari, H. (eds.) BPM 2018. LNBIP, vol. 342, pp. 361–373. Springer, Cham (2019). https://doi.org/10.1007/978-3-030-11641-5_29

29. Viriyasitavat, W., Xu, L.D., Bi, Z., Sapsomboon, A.: Blockchain-based business process management (BPM) framework for service composition in Industry 4.0. J. Intell. Manuf. **31**(7), 1737–1748 (2020)

30. Viriyasitavat, W., Xu, L.D., Niyato, D., Bi, Z., Hoonsopon, D.: Applications of blockchain in business processes: a comprehensive review. IEEE Access **10**, 118900–118925 (2022)

31. Weber, I., Xu, X., Riveret, R., Governatori, G., Ponomarev, A., Mendling, J.: Untrusted business process monitoring and execution using blockchain. In: La Rosa, M., Loos, P., Pastor, O. (eds.) BPM 2016. LNCS, vol. 9850, pp. 329–347. Springer, Cham (2016). https://doi.org/10.1007/978-3-319-45348-4_19

32. Weske, M.: Business Process Management - Concepts, Languages, Architectures, 3rd edn. Springer, Cham (2019). https://doi.org/10.1007/978-3-642-28616-2

33. Xu, Y., Slaats, T., Düdder, B., Debois, S., Wu, H.: Distributed and adversarial resistant workflow execution on the Algorand blockchain. CoRR abs/2211.08695 (2022)

34. Zhang, D., Xu, X., Zhu, L., Paik, H.: A process adaptation framework for blockchain-based supply chain management. In: Proceedings of International Conference on Blockchain and Cryptocurrency (ICBC), pp. 1–9. IEEE (2021)

35. Lim, Y.Z., Zhou, J., Saerbeck, M.: Shaping blockchain technology for securing supply chains. In: Zhou, J., et al. (eds.) ACNS 2021. LNCS, vol. 12809, pp. 3–18. Springer, Cham (2021). https://doi.org/10.1007/978-3-030-81645-2_1

Towards an Understanding of Trade-Offs Between Blockchain and Alternative Technologies for Inter-organizational Business Process Enactment

Martin Kjäer[(✉)][iD], Thomas Preindl[iD], and Wolfgang Kastner[iD]

TU Wien, Karlsplatz 13, 1040 Wien, Austria
{martin.kjaer,thomas.preindl,wolfgang.kastner}@tuwien.ac.at

Abstract. Several studies have proposed the application of blockchains in Inter-organizational Business Processes (IOBPs), primarily citing the technology's immutability, trust, and transparency as motivating factors. However, there is a notable lack of detailed comparisons between traditional, non-blockchain-based architectures and those incorporating this new technology. Such a comparison is critical for practitioners like software architects to fully comprehend blockchain-based solutions' strengths and their potential trade-offs and suitable scenarios for alternative technologies. This paper endeavors to bridge this knowledge gap by contrasting the attributes of public and private blockchains with those of Trusted Third Parties (TTPs) and Electronic Data Interchange (EDI) – the latter being a widespread method for automated data exchange between organizations. We underscore less explored advantages of blockchains, such as the ability to provide non-equivocation. Conversely, we identify that TTPs offers lower complexity levels and superior flexibility.

Keywords: BPM · blockchain · trusted third party · EDI · architectural concerns · trade-off analysis

1 Introduction

IOBPs have been affected by digital disruption for decades. First, by introducing EDI and proprietary communication networks, and later by the Internet in combination with modern Business Process Management Systems (BPMS). In recent years, the use of blockchains[1] has been considered and several publications in the business process management community have proposed novel architectures which make use of them.

Due to their nature, public blockchains provide some unique qualities which are hard to achieve otherwise. Overarching is the ability to provide a trustable

---

[1] Blockchains are a particular type of Distributed Ledger Technology (DLT). For the sake of simplicity, we will only use the term 'blockchain'.

J. Köpke et al. (Eds.): BPM 2023, LNBIP 491, pp. 36–50, 2023.
https://doi.org/10.1007/978-3-031-43433-4_3

state machine, whereby trust is achieved through a high degree of decentralization and introducing an incentive-aligned system, ensuring that rational actors behave as expected [17]. Until recently, for use cases requiring the highest level of trust, the use of a TTP was the best possible solution. As the name suggests, TTPs are chosen when a neutral and reliable entity is needed, which behaves [...] *in a well-defined way that does not violate agreed-upon rules, policies, or legal clauses* [...] [11, p. 7].

Measuring trust is, however, not easy and even if various methods have been developed, they are mainly assessed with the help of interviews and are also hardly used in practice [14]. Trust in an external system can be enhanced through various techniques such as audits, monitoring, and a comprehensible governance structure. Nevertheless, Singer and Bishop [19] go a step further and argue that trust should be considered harmful in the first place and thus minimized wherever possible. Blockchains cannot solve the trust problem per se, as humans will need to interact with them. As long as humans are responsible for signing certain transactions to trigger some action on-chain, the risks connected to the off-chain protection mechanisms that prevent the user from getting tricked into signing something unintentional remains.

However, when employed correctly, blockchains do have the potential to enhance specific guarantees connected to trust. One of them is non-equivocation, which is essential for use cases that require a high level of security [22]. An example where equivocation is problematic is the misuse of trust in certificate authorities (CAs). These have the power to distribute different certificates pointing to the same domain, which happened, e.g., to Google before [2]. Blockchains can solve this issue, as shown by Tomescu and Devadas [22].

Blockchains also provide benefits in the field of IOBPs as they allow trustless execution without the need for a trusted intermediate. However, when designing a system for IOBP enactment, deciding whether to include a blockchain within the BPMS architecture remains challenging. Several sources propose the use of blockchains in this domain when trustability, transparency, or immutability is required [24, Chapter 8], [13]. While these quality attributes are undoubted, we argue that deciding whether to use a blockchain in a BPMS also depends on many other aspects and requires the careful analysis of trade-offs between several concerns.

This work aims to move these concerns to center stage and to contribute towards a common understanding of the influencing factors and how they relate. Our contribution thus aims to provide value to software architects and assist them in making better design decisions. Furthermore, this work adds to research direction 5 of Mendling et al. on opportunities and challenges in this domain [15, p. 4:13]: *"Developing techniques for identifying, discovering, and analyzing relevant processes for adopting blockchain technology. Researchers will have to investigate which characteristics of blockchain as a technology best meet the requirements of specific processes."*

1.1 Related Work

Van der Aalst [23] contributes to understanding interoperability in IOBPs, mainly focusing on the concept of *capacity sharing*. Stiehle and Weber [20] provide a comprehensive view of business process enactment using blockchains. They introduce a taxonomy that distinguishes between supported capabilities and enforced guarantees, enriching the understanding of enactment facets. Also, Garcia-Garcia et al. [9] conducted a systematic literature review on collaborative business process management using blockchain.

Multiple authors have explored challenges and opportunities in this domain. Mendling et al. [15] provide a broad overview related to the use of blockchains, while Breu et al. [5]. specify four types of general challenges: Flexibility, correctness, traceability, and scalability. Architectures and threat mitigation are other aspects covered in the literature. Ordoñez-Guerrero et al. [16] conduct a systematic mapping study on architectural concerns, and Colwill [8] delves into the concern of insider threats, highlighting the limits of technical mitigation. Trust, a crucial element in this field, is examined by McEvily et al. [14], who summarize different trust measures and their practical application.

1.2 Chosen Approach

The set of possible solutions that aim to tackle challenges in IOBP enactment is significant. Therefore, to explore the trade-offs between blockchain-based architectures and their alternatives, we categorize them into distinct technology groups for comparison. To facilitate a comparison between these groups, we employ architectural concerns, which we subdivide into three categories:

1. Challenges in IOBPs that have been defined by Breu et al. [5]
2. Concerns that are related to blockchain-based architectures
3. Concerns related to TTPs

The challenges related to IOBPs as defined in [5] are flexibility, correctness, traceability, and scalability. Even for traditional technologies such as EDI, these are not easy to address; therefore, we argue that including them in our analysis adds value. Concerns related to blockchain-based architectures are mainly connected to the drawbacks of public blockchains. We choose to include privacy, transaction costs, and finality in our analysis, as these are well-known issues that are also not easy to address [20]. Finally, we include non-equivocation and insider threat prevention, concerns related to TTPs. We include them to foster the discussion of trade-offs connected to the main competitor (TTPs).

The rest of this work is structured as follows: The succeeding section includes a brief overview of EDI, a technology group focused on automating electronic business document exchange. EDI is the technology that is used when no TTP is employed for enactment of IOBPs. In Sect. 3, we briefly describe the other groups of technologies which tackle the challenges mentioned above. In Sect. 4, we discuss the described architectural concerns and trade-offs between technology groups. In the discussion (Sect. 5), we summarize our findings and briefly discuss the next steps.

2   EDI as Alternative to TTPs and Blockchains

Besides the challenges that TTPs and the different types of blockchains aim to address, EDI has enabled the automation of IOBPs for decades. We include this well-established technology group in our work for completeness, mainly because they provide some interesting, sometimes overlooked capabilities. In particular, when considering the use of blockchains for IOBP, we argue that in the first instance, however, it should be carefully evaluated whether the features of a TTP or an EDI-based system are not sufficient. This section briefly discusses EDI and which concerns it can tackle.

Due to various standardization efforts, EDI formats such as UN/EDIFACT[2] or UBL[3] are widely used and allow companies to exchange business documents such as purchase agreements or invoices in an electronic and automated manner. For communication, different messaging standards exist to ensure the secure exchange of business documents whereby several techniques such as message integrity verification, digital signatures, and Public Key Infrastructure (PKI) are used: Message integrity verification provides the ability to verify that electronic messages containing business documents have not been changed during transmission or when stored.

This is guaranteed by using cryptographic hash functions, which are a means to calculate the hash of a message before digitally signing it using electronic signatures. Given that the identity of the sender is undeniably connected to the public key,[4] the receiver can verify that received message has not been tampered with and originates from the expected party. This feature is comparable to transactions on a blockchain, which are also identified by their cryptographic hash value. However, compared to blockchains, where each participant can verify that a transaction has not been modified, EDI uses the concept of messages being exchanged only between individual participants.

Non-repudiation of Origin. Besides the capability of message integrity verification as well as sender authentication, the described mechanisms also provide another guarantee: The content of electronic business documents becomes non-repudiable. This property is also known as *non-repudiation of origin* and allows the receiver of a message in case of a dispute with the sender to prove evidence to a third party that a sender transmitted a particular message [25]. This guarantee makes another business partner liable for the actions taken and does not require any more advanced technology other than the described ones.

---

[2] https://unece.org/trade/uncefact/introducing-unedifact.

[3] https://www.oasis-open.org/committees/tc_home.php?wg_abbrev=ubl.

[4] This is usually solved by making use of PKI, but also other approaches such as Decentralized Identifiers (DIDs) exist.

3 Technology Groups for Inter-organizational Business Process Enactment

Technologies for secure IOBP enactment can be grouped into several categories. In this section, we present how we group them together with an overview of these groups. We include public- and private chains, commonly used as differentiation in the blockchain domain. To consider the growing sector of scaling solutions for blockchains, we also include so-called *Layer 2* solutions. TTPs have been added as they remain the main competitor of blockchains due to their trust offering.

3.1 Public Blockchains

Public blockchains such as Bitcoin or Ethereum offer publicly viewable data structures and trustless execution of transactions. As their name suggests, they are a public good and allow anyone to interact with them if a transaction fee is paid for their use. We follow a blockchain-agnostic approach – instead of discussing specific properties of blockchains (Proof-of-Work vs. Proof-of-Stake, etc.), we assume that the following non-functional properties, which have been described by Xu et al. [24] hold: *non-repudiation, integrity, transparency* and *immutability* of stored data as well as *equal rights*, which describes that every actor that wishes to access or manipulate the blockchain is allowed to do so.

Furthermore, we assume the following properties to be given:

- Programmability, i.e., the possibility to define smart contracts, which are executed on-chain[5]
- High degree of decentralization, nodes are distributed across many actors and nation-states
- High economic security so that fraudulent behavior such as double spending[6] becomes economically unattractive

3.2 Layer 2 Blockchain Solutions

Scalability of public blockchains is a challenge, which is not easy to solve, at least when the decentralization and security properties should be noticed [10]. One possibility to tackle this issue is the use of layer 2 solutions. Development in this segment has steadily progressed and several designs have been tried. The most promising ones are rollups, which maintain their own state machines but aim to derive the security guarantees of their connected 'base' layer, i.e., a public blockchain. [21]. This is achieved by using the base layer as a data availability layer, which stores an ordered list of all layer 2 transactions without executing them. With the transactions, rollup providers also publish a commitment of the updated rollup state. At this point, two different designs are used, which we briefly describe here.

---

[5] This property is only supported by second-generation blockchains such as Ethereum.

[6] In a double spending attack, a malicious actor attempts to spend owned tokens twice, e.g., by enforcing the re-ordering of already finished blocks.

Zero Knowledge Rollups. These types of rollups employ Zero Knowledge (ZK) proof schemes to calculate proofs of correct execution of new transactions. A smart contract deployed on layer 1 verifies this proof, which accepts it only if it is correct. The on-chain verification of state updates on layer 1 is the most critical aspect of this design, even if it requires higher complexity due to the use of ZK proofs. Due to the high costs of on-chain verification of ZK proofs, there is a trade-off between fast settlement and low transaction fees. Therefore sequencers only create proofs for batches of transactions to distribute the cost of verification.

Optimistic Rollups. In contrast to ZK rollups, optimistic rollups commit only the state root of the rollup, but no proof of correct execution. This design requires external actors to verify if a newly published state root is correct. They can do so by executing all new transactions before comparing the calculated new state root of the rollup with the claimed state root published on-chain. An external actor can publish a so-called fraud-proof in case of a deviation, which is then automatically verified on-chain [21]. If the fraud-proof is correct, the sequencer, the node that maintains the rollup, has to pay a penalty fee, and the state machine of the layer 2 solution gets *rolled back* to the point before the incident.

3.3 Private Blockchains

Private blockchains have emerged as an alternative design approach after public blockchains became popular. Compared to their public pendant, they can only be advanced by a group of known participants that maintain it, which is the reason why private blockchains are also named permissioned blockchains. The advantages of private blockchains are that only the chain's contributors have access, which is why the concept has been especially interesting for business use cases. Furthermore, the nodes of a private blockchain can use professional server infrastructure. That means that throughput is only limited by server expenses, and transaction fees can be kept low, which are further advantages for businesses.

3.4 Trusted Third Party

A TTP can act as an intermediary and coordinate processes between multiple parties. Interoperability is enabled through *capacity sharing* and several entities make use of a centralized workflow manager [23]. For example, some of the largest and most trusted entities, which execute financial workflows, so-called Real-Time Gross Settlement System (RTGS) platforms, can be named. These systems, which central banks typically operate, execute transactions between banks on behalf of their customers.[7] Besides these highly payments-specialized systems, many other systems exist for other use cases, ranging from generic trustable BPMS service providers to use-case-specific implementations.

---

[7] An example for an RTGS is *T2* of the European Central Bank (ECB): https://www.ecb.europa.eu/paym/target/.

4   Architectural Concerns

This chapter analyzes different architectural concerns and how they relate to the technologies discussed in the previous section. In Table 1, we include selected technology groups (in columns) as well as the selected architectural concerns (in rows). It is important to note that the assessment does not claim completeness. Instead, it should indicate whether a specific concern can be addressed better or worse by a given technology group and how that relates to other technology groups.

We base the estimation shown in Table 1 on our argumentation, which forms the remainder of this chapter. For each concern, the scaling is adjusted relatively between the individual technology groups between zero and four points. This means that each concern has at least one technology group with a value of zero points and one with a value of four points. The table also shows ranges (visualized with hatched circles), as in some cases, the ability to address an architectural concern depends heavily on specific design decisions. An example is privacy on public blockchains: most public blockchains do not offer privacy natively, so if no other arrangements are made, everyone has visibility of all transactions. Nevertheless, sophisticated privacy protection mechanisms exist that leave only small amounts of metadata on the blockchain. In this case, we show a range from zero (no privacy protection) to three (privacy protection, but metadata left on the blockchain).

4.1   Flexibility

Achieving flexibility in IOBPs is challenging and requires carefully designing systems capable of responding to extraordinary situations. An example of flexibility is, e.g., the ability to adapt a process instance due to an allergic reaction of a patient in a treatment process within a hospital [18, Chapter 3]. For processes that require strict enactment, such as enforcing the compliance to anti-money laundering provisions, flexibility needs to guarantee that changes to the process or a process instance follow specific rules so that the process is still compliant after the change [18, Chapter 10]. These examples show how complexity increases when flexibility needs to be implemented correctly.

For IOBPs that are enacted by utilizing a public blockchain, flexibility presents an even more complicated issue: As application code on public blockchains is by default immutable, flexibility is hard to achieve in that context. Any necessary changes in the behavior have to be foreseen at the development time and reconfiguration capabilities have to be explicitly included in the application.

The same holds for layer 2 solutions as these instances are also designed to operate as intended, so every form of flexibility must be provided at design time. Also, in the case of private chains, the same level of flexibility can be achieved, with one exception: If an agreement between the participating parties can be achieved, it becomes possible to update the blockchain, which means that even historic transactions could be changed retroactively.

Table 1. Architectural concerns and how different groups of technologies can address them. More filled circles mean that the technology group is more capable of tackling the concern. Additional hatched circles show a range, with the minimum being the last filled circle (if there is any) and the maximum being the last hatched circle.

Architectural Concern \ Technology Group	Public Blockchains	Layer 2 – ZK Rollups	Layer 2 – Optimistic Rollups	Private Blockchains	Trusted Third Parties (TTPs)	Electronic Data Interchange (EDI)
Concerns Related to IOBPs [5]						
Flexibility	⊘⊘○○	⊘⊘○○	⊘⊘○○	●●●○	●●●○	●●●●
Correctness	●●●●	●●●●	●●●●	●●●○	●●○○	○○○○
Traceability	●●●●	●●●●	●●●●	●●○○	●●○○	○○○○
Scalability	○○○○	●●○○	●●○○	●●●⊘	●●●●	●●●●
Concerns Related to Blockchains						
Privacy	⊘⊘⊘○	⊘⊘⊘○	⊘⊘⊘○	●●⊘○	●●●○	●●●●
Transaction Costs	○○○○	●●○○	●●○○	●●●○	●●●○	●●●●
Finality	●●○○	●○○○	⊘○○○	●●●○	●●●●	●●●●
Concerns Related to TTPs						
Non-equivocation	●●●●	●●●●	●●●●	●●●○	●●○○	○○○○
Insider Threat Prevention	●●●○	●●●○	●●●○	●●●○	●○○○	○○○○

In the case of TTPs, flexibility depends on the willingness and capability of the TTP. Since the unique selling point of a TTP is *literally* trust, all types of flexibility need to be defined upfront.

We summarize that all listed technology groups need to consider the flexibility of process enactment at design time, with public blockchains and layer 2 solutions being the technology groups that are the least forgiving if flexibility is not properly considered during design time. An exception to this statement are EDI solutions, as they provide only the data exchange layer between the local BPMS and the external actors. This setup allows to adapt flexibility internally, which cannot be achieved by the other technology groups.

4.2 Correctness

Achieving correctness of IOBP enactment is another challenge that is hard to tackle, especially when many parties are involved in a process [5]. In the case of public blockchains, the correct enactment of an IOBP can be guaranteed, given that the process is correctly implemented on-chain. Transactions sent to the blockchain advance the state of the process engine on-chain and are only accepted if they follow pre-defined rules. Adapting a process that stores its state on-chain becomes possible only if governance mechanisms or alternative explicit possibilities have been implemented to enable such a change. This presents, however, also a correct evolvement, as these kinds of capabilities have to be provided before deployment.

Layer 2 solutions also provide correct execution guarantees, as only valid layer 2 transactions are accepted on layer 1 (in the case of ZK-rollups) or eventually finalized (in the case of optimistic rollups). That means that correct enactment of an IOBP, which uses a layer 2 solution, can be assured. Private blockchains can enforce correctness within their closed group of participants. This means they provide lower guarantees than public blockchains, which do not restrict their set of participating nodes. Compared to private blockchains, the correctness of process enactment on a TTPs depends on the capabilities of this single external party. Correctness, therefore, indirectly relates to the level of trust associated with the TTP. In the case of EDI solutions, correctness can only partly be guaranteed, e.g., with non-repudiation of origin. Other guarantees, such as *non-repudiation of receipt* or *non-repudiation of delivery*, require a third external party [25].

4.3 Traceability

Due to the distributed enactment of IOBPs, traceability presents a subsequent stumbling block, as processes are spanned across many companies. Not every party of a more extensive process receives and sees all necessary events [5]. Since public blockchains offer the possibility that anyone can investigate all transactions, they provide a suitable answer to this problem.[8] As the ledger is also immutable, public blockchains guarantee that process traceability is also provided retrospectively for parties not involved during process enactment.

Layer 2 solutions provide the same properties, as their transactions are accessible to anyone, finalized on a public blockchain and therefore also offer immutability. Compared to that, private blockchains cannot grant the same level of traceability, as transactions are only shared between the approved participants that maintain it. This means that traceability is only offered to these parties, while external actors must trust the consortium. TTPs are very similar in this respect, as they make data available to all parties with access rights. Similarly, like with correctness, EDI solutions offer only limited possibilities due to the abovementioned reasons.

4.4 Scalability

Public blockchains, which follow the principle of maximal decentralization, deliberately restrict transaction throughput whereby computational resources are spared. This maxim allows more participants to operate a blockchain node that

---

[8] At this point, we neglect the concern of privacy, which is discussed in Sect. 4.5.

verifies and stores all blockchain transactions and allows increased decentralization and higher security. Obviously, for enacting IOBPs on public blockchains, this presents a significant obstacle since only the use cases that can operate under these limited resources can be implemented.[9]

In contrast, layer 2 solutions are, at the time of writing, able to achieve a throughput that is approximately two orders of magnitude higher than their base layer [21]. This represents a significant advantage compared to public chains, even if scalability is still limited (approx. 3000 transactions per second [21]). It is important to note that this figure refers to the throughput of only one rollup, with multiple rollups in operation on a public blockchain. It remains to be seen if these metrics can be further increased, as research in this field is only a few years old. In contrast, private blockchains, TTPs and EDI solutions do not share the issue of artificially restricted throughput with their competitors. Scalability is thus only restricted by the limits of hardware and software components but not by other factors, representing a significant advantage over the other technology groups.

4.5   Privacy

Privacy is a necessary pre-condition for almost all electronic business communication use cases. These range from the desire of companies not to disclose their purchase prices to legal prohibitions concerning antitrust law [12]. Most public chains do not natively support privacy, as their ledger of transactions is not encrypted. Privacy-preserving solutions such as zero-knowledge proof schemes,[10] which can be built on top of them have the potential to tackle this problem. However, they also have drawbacks, as they demand higher computational resources and imply higher complexity levels. Research on this topic is ongoing and mainly focused on increasing performance [7].

As their name suggests, private chains allow better protection for privacy, as the blockchain itself is not publicly accessible. This increases privacy substantially, but the participants that maintain the blockchain can still view the transactions of all parties. For most companies, this is an issue they cannot accept, which is why different designs have been proposed to overcome this issue by establishing an additional private environment between actors that is only linked to the private chain [6].

TTPs provide a high level of privacy, as data is only shared with parties that have been granted access explicitly and EDI solutions allow even higher levels of privacy since information is only shared between the parties that necessarily need the information (compared to TTPs).

---

[9] Aside from the risk of inducing higher transaction costs if the application requires a substantial part of these resources.

[10] It is essential to note that privacy based on zero knowledge should not be confused with zero knowledge *rollups*, which usually do not provide privacy, even if they are based on the same group of technologies.

4.6   Transaction Costs

Transaction costs can be a significant drawback for using IOBP enactment on public blockchains. As previously shown, some blockchains have become very expensive and fee prices can be highly variable [20], both serious business issues.

In contrast, private blockchains and TTPs are substantially better, as they are only limited by computational and infrastructure costs. Private blockchains have a disadvantage compared to TTPs, as several instances need to operate computational infrastructure, whereas a TTP only needs to host one instance. Nevertheless, for use cases requiring the highest security standards, the costs of a TTP might be considerable due to the system's complexity and security requirements. E.g., the previously mentioned Real Time Gross Settlement (RTGS) system of the European Central Bank charges up to EUR 0.8 per transaction, which is an amount that is much higher than any other high-volume transactional [3]. This example highlights the importance of the required security level, which impacts the willingness to pay higher transaction fees.

4.7   Finality

Reaching transaction finality presents an issue connected to distributed systems, especially public blockchains. Finality in the context of blockchains describes the point in time when a previously proposed block cannot be removed from the canonical chain of blocks anymore. Different blockchains offer different types of finality (deterministic or probabilistic [4]), but the issue cannot be removed completely, as the need to reach consensus remains.

This issue is worse for optimistic rollups, as the finality can only be reached after a longer time, during which fraud proofs can be submitted [21]. Therefore, this more extended time period presents an intentional design decision necessary for the system to function. The only possibility to lower this time period is if a full copy of the optimistic rollup is maintained and verified locally. The independent off-chain execution of all transactions provides the ability to verify the correctness of the updated state root and thus to identify de-facto finality before fraud proofs can be submitted.

ZK rollups, in contrast, provide finality once the ZK proof gets published on the layer 1 blockchain. This provides an advantage compared to optimistic rollups Independent from the specific implementation, all rollups lag behind their connected layer 1 blockchains and thus require longer to reach finality.

Lower finality times can only be offered by all other technology groups (private blockchains, TTPs and EDI solutions). Private chains have a minor disadvantage compared to the other two technologies, as they also need to reach consensus within the set of nodes.

4.8   Non-equivocation

Due to their logically centralized yet organizationally decentralized data structures [1], public blockchains offer good protection against equivocation. ZK rollups provide similar guarantees, as transactions are verified on layer 1 once the proof is submitted. This contrasts optimistic rollups, where state updates are published optimistically and opens a short time window (until a fraud proof gets published), where equivocation would theoretically be possible.

Private chains can also be considered to tackle this issue, as equivocation would require that the participating nodes collude. Nevertheless, collusion is easier to achieve in private blockchains than in an architecture that is based on a public blockchain. Compared to that TTPs don't require any form of consensus between multiple nodes when storing and distributing data. Unfortunately that also makes equivocation easier, as the TTP could deliver different versions of the data to different participants. To prevent such a Byzantine behaviour, participants could gossip messages received from the TTP in order to detect this type of attack. Nevertheless, these kind of techniques require significant additional networking and are not easy to apply in practice [22]. EDI message exchange protocols don't provide non-equivocation protection natively, so any more advanced capabilities such as consensus or gossiping of exchanged messages would need to be added on the application layer.

4.9   Insider Threat Prevention

Insider threats are a challenge related to TTPs that is not easy to overcome. They can partly be mitigated by employing technical control mechanisms such as access control or allowing only minimum privilege [8]. All technically possible security precautions should be applied, especially in areas where the highest security standards are required due to potentially disastrous outcomes in case of an insider attack. Public and private blockchains, as well as layer 2 blockchains, all provide some possibilities to defend against insider threats: Everything encoded on-chain (in smart contracts or enshrined in the blockchain protocol) will be executed as defined. This capability provides the parties that are interacting with an application hosted on-chain with the guarantee that operations that are not permitted on-chain will not be executed. However, it is still necessary to employ high security controls such as minimal access to all off-chain components. Blockchains (public and private) and layer 2 blockchains cannot mitigate these risks.

Compared to blockchain-based architectures, TTPs have fewer capabilities to protect themselves against insider threats. As they act as a single source of truth, a malicious insider can, if all other security precautions are overcome, manipulate information. At the same time, external parties may not be able to notice it. Depending on the use case, these kinds of manipulations may pose a severe risk. EDI solutions do not provide protection mechanisms against insider threats, as they are usually not the technology of choice for use cases that require high security between several businesses (where TTPs are usually chosen).

5   Discussion

Our analysis sheds light on various architectural concerns related to technology groups in the field of IOBP enactment. We identified several implications associated with certain architectures. One of them is that a BPMS that makes use of public blockchains requires the design of systems with higher levels of complexity, e.g., to address the need of flexibility as this presents an issue that can't be added after on-chain deployment. The same holds true for layer 2 solutions, which at least offer some relief in regards to transaction costs and scalability. Nevertheless this novel technology also induces higher complexity and dependencies, as new components (rollups) as well as additional actors (sequencers) have to be added.

One of the objectives of this work is to provide software architects with a means to identify trade-offs between different technology groups in the area of IOBP enactment. Table 1 summarizes our findings and offers a first step towards this vision. Our approach does not only cover well-known drawbacks of blockchain-based solutions, such as privacy or transaction costs, but also concerns related to TTPs, such as insider risks or the risk of equivocation. We argue that our approach enriches the discussion by placing the emphasis on these specific capabilities which are only indirectly related to the more generic quality attribute *trust*.

In addition to the architectural concerns discussed, practical aspects such as technology maturity must also be considered when choosing the most appropriate architecture. EDI solutions and TTPs are the most mature technology groups in our comparison, as a set of EDI standards and also well-established TTPs exist. On the other hand, e.g., ZK- or optimistic rollups are still relatively early, even if they might hold great future potential. Both public- and private blockchains fall in between, although they are more early-stage in maturity.

Finally, we argue that conventional technologies such as EDI should be moved closer to the center of the discussion of blockchain-based architecture for IOBP enactment. These technologies are well-established in the industry and will likely be around for longer. Including them in the discussion adds value, especially for industry experts and practitioners.

Limitations and Future Work. Our approach does not claim completeness about the selected architectural concerns. Also, the estimation of how well these architectural concerns can be addressed by the discussed technology groups (Table 1) is not based on metrics. However, we plan to include metrics related to concerns in future works.

Furthermore, some solutions that tackle challenges in this domain do not fit in our technology groups (e.g., if only commitments of an off-chain BPMS are published on a public blockchain). However, considering all possible architectures is beyond the scope of this work and provides an opportunity to expand upon it in future contributions. We also plan to extend our approach by including further architecturally relevant concerns such as composability or service uptime.

6   Conclusion

Enactment of IOBPs is challenging due to several influencing factors, ranging from a lack of standardization and automation to technical limitations that prevent adoption. Various literature sources argue that a need for increased transparency, trust, and immutability are reasons for employing blockchains in this field. Our analysis of architectural concerns and trade-offs between them goes beyond the state of the art by including concerns related to the field of IOBPs as well as concerns connected to specific technologies that aim to tackle some of the challenges of this domain. We also move traditional systems such as TTPs and EDI into the spotlight and compare them against blockchain-based solutions. We identify several critical issues, such as the need for complex setups when a certain level of flexibility is required in a blockchain-based architecture. Overall, our work is the first step towards a more holistic approach to architectural concerns in this domain.

Acknowledgements. This research has been partially supported and funded by the Austrian Research Promotion Agency (FFG) for the research project "DiCYCLE – Reconsidering digital deconstruction, reuse and recycle processes using BIM and Blockchain" under the contract number 886960.

References

1. Bitcoin: Clarifying the foundational innovation of the blockchain. https://continuations.com/post/105272022635/bitcoin-clarifying-the-foundational-innovation-of. Accessed 04 May 2023
2. Further improving digital certificate security. https://security.googleblog.com/2013/12/further-improving-digital-certificate.html. Accessed 4 May 2023
3. TARGET services pricing guide. https://www.ecb.europa.eu/paym/target/consolidation/profuse/shared/pdf/ecb.targetservicespricingguide_v1.0.en.pdf. Accessed 04 May 2023
4. Anceaume, E., Del Pozzo, A., Rieutord, T., Tucci-Piergiovanni, S.: On finality in blockchains. In: 25th International Conference on Principles of Distributed Systems (OPODIS 2021). Dagstuhl Publishing (2022)
5. Breu, R., et al.: Towards living inter-organizational processes. In: 2013 IEEE 15th Conference on Business Informatics, pp. 363–366 (2013)
6. Brotsis, S., Kolokotronis, N., Limniotis, K., Bendiab, G., Shiaeles, S.: On the security and privacy of hyperledger fabric: challenges and open issues. In: 2020 IEEE World Congress on Services (SERVICES), pp. 197–204 (2020)
7. Capko, D., Vukmirovic, S., Nedic, N.: State of the art of zero-knowledge proofs in blockchain. In: 2022 30th Telecommunications Forum (TELFOR). IEEE (2022)
8. Colwill, C.: Human factors in information security: the insider threat - who can you trust these days? Inf. Secur. Tech. Rep. **14**(4), 186–196 (2009)
9. Garcia-Garcia, J.A., Sánchez-Gómez, N., Lizcano, D., Escalona, M.J., Wojdyński, T.: Using blockchain to improve collaborative business process management: systematic literature review. IEEE Access **8**, 142312–142336 (2020)
10. Halpin, H.: Deconstructing the decentralization trilemma. In: Proceedings of the 17th International Joint Conference on e-Business and Telecommunications. SCITEPRESS (2020)

11. Baseline identity management terms and definitions: Standard. ITU-T, Geneva, CH (2021)

12. Kim, K., Justl, J.M.: Potential antitrust risks in the development and use of blockchain. J. Taxation Regul. Finan. Inst. **31**(3), 5–16 (2018)

13. Ladleif, J., Weske, M., Weber, I.: Modeling and enforcing blockchain-based choreographies. In: Hildebrandt, T., van Dongen, B.F., Röglinger, M., Mendling, J. (eds.) BPM 2019. LNCS, vol. 11675, pp. 69–85. Springer, Cham (2019). https://doi.org/10.1007/978-3-030-26619-6_7

14. McEvily, B., Tortoriello, M.: Measuring trust in organisational research: review and recommendations. J. Trust Res. **1**(1), 23–63 (2011)

15. Mendling, J., et al.: Blockchains for business process management - challenges and opportunities. ACM Trans. Manage. Inf. Syst. **9**, 1–16 (2018)

16. Ordoñez-Guerrero, A.C., Muñoz-Garzon, J.D., Roberto Dulce Villarreal, E., Bandi, A., Ariel Hurtado, J.: Blockchain architectural concerns: a systematic mapping study. In: 2022 IEEE 19th International Conference on Software Architecture Companion (ICSA-C), pp. 183–192 (2022)

17. Pass, R., Seeman, L., Shelat, A.: Analysis of the blockchain protocol in asynchronous networks. In: Coron, J.-S., Nielsen, J.B. (eds.) EUROCRYPT 2017. LNCS, vol. 10211, pp. 643–673. Springer, Cham (2017). https://doi.org/10.1007/978-3-319-56614-6_22

18. Reichert, M., Weber, B.: Enabling Flexibility in Process-Aware Information Systems. Springer, Heidelberg (2012). https://doi.org/10.1007/978-3-642-30409-5

19. Singer, A., Bishop, M.: Trust-based security; or, trust considered harmful. In: New Security Paradigms Workshop 2020, pp. 76–89. ACM (2021)

20. Stiehle, F., Weber, I.: Blockchain for business process enactment: a taxonomy and systematic literature review. In: Marrella, A., et al. (eds.) BPM 2022. LNBIP, vol. 459, pp. 5–20. Springer, Cham (2022). https://doi.org/10.1007/978-3-031-16168-1_1

21. Thibault, L.T., Sarry, T., Hafid, A.S.: Blockchain scaling using rollups: a comprehensive survey. IEEE Access **10**, 93039–93054 (2022)

22. Tomescu, A., Devadas, S.: Catena: Efficient non-equivocation via bitcoin. In: 2017 IEEE Symposium on Security and Privacy (SP), pp. 393–409 (2017)

23. van der Aalst, W.M.: Process-oriented architectures for electronic commerce and interorganizational workflow. Inf. Syst. **24**(8), 639–671 (1999)

24. Xu, X., Weber, I., Staples, M.: Architecture for Blockchain Applications. Springer, Cham (2019). https://doi.org/10.1007/978-3-030-03035-3

25. Zhou, J., Gollmann, D.: Evidence and non-repudiation. J. Netw. Comput. Appl. **20**(3), 267–281 (1997)

Towards Object-Centric Process Mining for Blockchain Applications

Richard Hobeck[1]([envelope]) and Ingo Weber[2]([envelope])

[1] Chair of Software and Business Engineering, Technische Universitaet Berlin,
Berlin, Germany
`richard.hobeck@tu-berlin.de`
[2] Technical University of Munich, School of CIT, and Fraunhofer Gesellschaft,
Munich, Germany
`ingo.weber@tum.de`

Abstract. Process mining event logs are traditionally formatted to reflect the execution of a collection of individual process instances, with a fixed case notion. In practice, process instances are often intertwined, and the scope of a particular process is less static. When flattening complex application data to traditional event log formats, like XES, problems such as divergence and convergence occur in the resulting event logs. Object-centric process mining with its object-centric event log (OCEL) format were introduced to tackle these issues, supporting multiple case notions with a single event log. While the adoption of object-centric logging is starting to gain momentum in several domains, the use of OCEL for event logs of blockchain applications has seen little research. In this paper, we investigate blockchain data structures and map them to OCEL logging capabilities. We present an approach to extracting data from blockchain applications that requires little domain knowledge. We discuss how to map data items to fit object-centric event logs and provide and analyze a corresponding OCEL event log for a blockchain application. The approach is evaluated based on a use case and contrasted to a previous case study.

Keywords: process mining · OCEL · blockchain

1 Introduction

Blockchain is a distributed ledger that allows for computer-based coordination between parties in an otherwise trustless environment [21, p. 3]. Second generation blockchains (e.g., Ethereum) can act as execution engines for arbitrary code. Applications that are implemented on blockchains (*decentralized applications*, short: DApps) generate data during runtime that can be made subject to data analysis. However, blockchain data may be fragmented and encrypted, and accounts and keys may change over time: properties that pose challenges to analyzing blockchain data, e.g., with process mining techniques [14].

Process mining is a set of techniques to generate knowledge from process data [10]. As process mining is heavily dependent on data, different data

© The Author(s), under exclusive license to Springer Nature Switzerland AG 2023
J. Köpke et al. (Eds.): BPM 2023, LNBIP 491, pp. 51–65, 2023.
https://doi.org/10.1007/978-3-031-43433-4_4

exchange formats have been discussed to achieve interoperability between data producing and data consuming systems. Widely-adopted as a format for event logs and an input format for process mining tools is the *eXtensible Event Stream* (XES) thanks to an early proclamation as an event log format standard [8]. XES allows only a single case notion which can cause problems in the event log, e.g., divergence and convergence [3]. To overcome these problems, the more recently established *Object-Centric Event Log* (OCEL) standard allows multiple case notions in one event log [7].

To date, XES remains the most widely-used standard for process mining event logs for blockchain applications [4]. Several approaches to creating XES event logs from blockchain data for process mining have been proposed. An overview of the existing approaches was presented in a recent systematic literature review [13]. [11] focuses on creating single-case notion event logs from blockchain log entries that are specified in smart contract code and written onto the blockchain. Their tool has been developed further and made blockchain platform independent [5]. [6] used fine-grain execution data of transactions that were generated by Ethereum nodes. They grouped the data by their sender address to create a case notion. Similarly, [16] decoded smart contract function calls and used names of called functions as activity names in event logs. [18] introduced steps to analyze the sojourn time between submissions of transactions and their inclusion in a block. There have also been initiatives to make blockchain event logs less dependent on a single-case notion. [12] extracted log entries for the Ethereum DApp *Cryptokitties* and presented a "artifact-centric" logging format based on an OCEL extension. The extension, however, has no process mining tool support. Current tools that create XES logs from blockchain data attempt to deal with data fragmentation and account changes by limiting the logging to certain accounts and event types impeding to document the full extent of a DApp's behavior.

In this paper, we address the above limitations with a two-pronged approach, specifically object-centric process mining for DApps and more comprehensive data capture. As such, we explore using the OCEL logging format to incorporate various types of execution data of an application in a blockchain environment in a single event log. The approach is aimed at suiting dynamic deployments and distributed logging, as commonly seen in blockchain applications. We make the following contributions: (1) we propose an object-centric approach to retrieving and decoding blockchain application data, (2) we broaden the set of considered data sources to include, among others, function calls and input data, logged variable values, dynamically created smart contracts and application structure, and tracking digital assets including cryptocurrency and (non-)fungible tokens (FTs and NFTs); (3) we conduct an initial evaluation of the proposed approach to investigate technical feasibility and comparatively analyze the resulting OCEL-log with an earlier case study that was based on XES.

We argue that challenges in creating event logs for blockchain applications can effectively be tackled using object-centric logging formats. In this paper, we explore particularities of data structures in a blockchain environment and map them to the capabilities of OCEL. We describe a data extraction method that flexibly responds to changes in a DApp's structure and updates of application code. For an initial validation of the approach, we conduct an assertion [22] as a basis for future experiments.

The remainder of the paper is structured as follows: In Sect. 2 we describe the XES and OCEL logging formats. Sect. 3 describes data attributes that relate to execution data in a blockchain environment (specifically Ethereum) and that may be part of a comprehensive event log of a DApp. Our approach to extracting blockchain data is presented in Sect. 4. This is succeeded by presenting extracted data for the DApp *Augur*, and analysis and an evaluation of the approach in Sect. 5. The evaluation compares data availability and process mining insights with a case study performed on the same DApp with a single-notion event log. We discuss the findings of the paper in Sect. 6 and conclude the paper in Sect. 7.

2   Logging Formats in Process Mining

For every process mining endeavor, which aims to derive insights from process data, the data is the foundation. To ensure interoperability between different information systems (e.g., systems producing event data as output and systems consuming event data as input for process mining) a standardized exchange format called *eXtensible Event Stream* (XES) was agreed on in 2010 and defined as a standard in 2016 [1,2]. XES is an instantiation or dialect of the eXtensible Markup Language (XML). XES defines syntax and semantics for an event log. As its main ingredients, XES defines the notions of a log, traces, events, and their respective attributes [2]. A log can contain traces, which in turn comprise events. A trace represents a *single* process instance.

However, during actual process executions, events often belong to multiple cases. An often-cited example is that of an online store where a customer created multiple orders with multiple items each; the store might send all items in a single shipment, or each item separately; and the customer might receive one invoice, or one per order, or one per shipment. What should the case notion be based on: one case per shipment, order, invoice, or customer? For most answers, cases are intertwined. Deciding on a case notion and creating a corresponding XES log hence includes a step of *flattening the log*. Approaches to flatten the event log may result in some of the following three issues. (1) *Deficiency*: if an event does not have the chosen case notion it does not appear in the data. (2) *Convergence*: if an event relates to two distinct cases in the chosen case notion, it will appear duplicated in the data. (3) *Divergence*: events of two different cases and a shared third case may appear causally related although they are not [1,3].

To address these problems a new standard was published in 2020: Object-centric Event Log (OCEL) [7]. Like XES, OCEL also comprises events and their attributes. A major innovation of OCEL allows events to refer to *multiple* case notions. This is achieved by introducing *objects* which represent "physical and informational entities composing business processes such as materials, documents, products, invoices, etc." [7, p. 4]. Objects and events have many-to-many relations in OCEL (see Fig. 1).

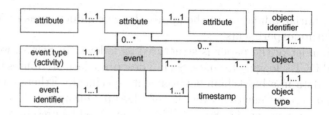

Fig. 1. UML diagram conceptualizing OCEL (adaption from [7]).

3 Data Attributes for Object-Centric Event Logs from the Ethereum Blockchain

Second-generation blockchains like Ethereum can be used as infrastructure for the execution of code that is deployed as smart contracts in *contract accounts* (CAs). Apart from CAs, a second type of accounts exists that are controlled by the private key holder (e.g., a user); these are called *externally owned accounts* (EOA)[1]. A blockchain is an append-only ledger. Data of the ledger is stored in consecutively created *blocks*. Each block is linked to its predecessor. Blocks contain *transactions* and *transaction receipts*, among others. Transactions capture transitions from one state of the blockchain to another [19]. The result of executing a transaction is hashed, and the hash is stored in the respective receipt; the receipt allows verifying that each machine that executes a transaction arrives at the same result, including CA log entries created as part of a contract invocation. Such log entries are used to communicate on-chain data as well as state changes to off-chain processing units.

On the Ethereum blockchain, transactions may contain invocations of smart contract functions, which result in computational steps being executed on the Ethereum Virtual Machine (EVM). These computational steps can be captured in transaction traces. Transaction traces exist on different levels of granularity. They can include assembly-level operations, e.g., comparisons and bit-wise logic operations, or push operations (within the EVM's computational stack); but they also comprise logging operations (emitting log entries) and system operations (e.g., creations of CAs and message calls between CAs) [19]. In contrast to earlier approaches, we here make use of relevant parts of traces. Fig. 2 is a class diagram depicting the relations between transactions, contracts, log entries, and traces.

Transaction traces exist temporarily and are not stored permanently in blocks of the Ethereum blockchain. In order to retrieve traces for historic transactions, transactions have to be replayed on the EVM (replay the transition from one state of the blockchain to the next) and the computational steps have to be stored separately from the blockchain. For that purpose, different tracers exist for the most widely used Ethereum execution client Geth[2,3].

---

[1] https://ethereum.org/en/developers/docs/accounts/, accessed 2023-05-15.
[2] https://ethereum.org/en/developers/docs/nodes-and-clients/, accessed 2023-06-03.
[3] https://geth.ethereum.org/docs/developers/evm-tracing/built-in-tracers, accessed 2023-05-16.

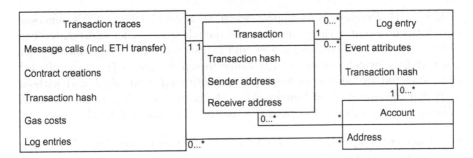

Fig. 2. Class diagram showing relations between contracts, transactions, events, and traces.

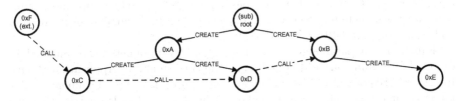

Fig. 3. Schema of a single tree-like deployment structure of smart contracts and possible message calls between accounts.

Depending on the objective of a **process analysis**, different data are relevant. In the context of user and contract behavior analysis as well as value stream analysis, from the sources above we considered the data listed in the following as relevant.

Contract Creations: Knowing the accounts belonging to a DApp is essential to identify transactions that concern the DApp and contain user or application data. The creation of a CA is documented in a transaction trace with the mnemonic "CREATE" or "CREATE2" combined with other data attributes (e.g., creator and costs of the creation). The EVM interprets the accompanying input data and attempts to deploy it as smart contract code. CAs can be added to a DApp at initial deployment, but there are also mechanisms for updating or adding smart contracts to running DApps, e.g., the factory pattern, registry pattern [20], or diamond proxy[4]. A set of CAs belonging to one DApp that gets deployed can be seen as a tree structure (or a forest consisting of several trees or sub-trees). Every CA of a DApp has a single creator (parent) and can have several children that it can create (e.g., through the aforementioned patterns). Additionally, the CAs of the DApp can send each other message calls laterally without traversing the creation branches. Contracts of the DApp can also be called from outside of the DApp (see Fig. 3).

Message Calls of CAs and EOAs: Message calls between accounts can be of different kinds, and are all documented in transaction traces with the mnemonic

---

[4] https://eips.ethereum.org/EIPS/eip-2535, accessed 2023-05-16.

"CALL". (1) *Function calls* appear when an account calls a CA's function. Data attributes sent with the message call include, e.g., sender address, receiver address, input and output data, fees, and transfers of Ethereum's native token Ether (ETH) (which can be 0). (2) *Ether transfers without function calls* are solely directed to value transfers without additional input data and without triggering the execution of function logic. Accompanying data are, e.g., sender address, receiver address, fees, and the amount of transferred ETH.

Log Entries: Log entries are used to communicate information about CAs' code execution to entities outside of the smart contract[5]. Log entries have to be specified in smart contract code in order to be emitted. Within limits, developers can choose what information shall be exposed with log entries. If smart contracts operate with ERC-tokens, however, developers are advised to implement standard interfaces including certain sets of events, e.g., to document token creation and transfers (e.g., ERC-20[6], ERC-721[7]).

Function call parameters as well as log entry parameters are emitted in encoded form and can be decoded using the corresponding Application Binary Interface (ABI) of a CA. ABIs are created at compile time and document a CA's interface functions (which can be called) as well as a set of log entries that are defined in the smart contract or inherited from other contracts[8]. An ABI's coverage of the log entries is not always complete. Before the recent introduction of Solidity v0.8.20, log entries according to the ERC-standard as well as log entries emitted through invoked code from imported libraries were not included in an ABI.

4   Data Extraction Method

The goal of the data extraction method is to gather as much data about a DApp as possible with minimum knowledge about the DApp. Therefore, the extraction method has to be capable of identifying DApp information from a small amount of input data. Hence, the extraction method takes as input a) *a non-empty set of accounts of a DApp*, and b) *a block range*. In order to extract the execution data of the DApp, two steps follow: 1) discover the creation sub-tree(s) of the DApp and 2) compute and transform execution data. Figure 4 visualizes the extraction method that is described in this chapter. Since the approach exploits information of transaction traces and partially decodes them, it can be seen as an extension of [16].

1) Discover the Creation Sub-tree(s) of a DApp. We start with the *input set of accounts of the DApp* and *a block range*. All operations will take place

---

[5] https://ethereum.org/en/developers/docs/smart-contracts/anatomy/#events-and-logs, accessed 2023-06-04.

[6] https://eips.ethereum.org/EIPS/eip-20#events, accessed 2023-05-16.

[7] https://eips.ethereum.org/EIPS/eip-721#specification, accessed 2023-05-16.

[8] https://docs.soliditylang.org/en/v0.8.13/abi-spec.html, accessed 2023-05-16.

within the specified block range. First, we retrieve transactions that have one of the input accounts as a sender or receiver account. We then look for transactions that contain message calls with ETH transfers from or to accounts from the input accounts. The newly identified transactions are replayed by an EVM of a Geth Ethereum Archival node to retrieve the full traces for the transactions. For that purpose, a build-in debug tracer is used with the call tracer setting and instructed to output log entries[9]. We then forward and backward search for DApp accounts along the create sub-tree to identify: a) Child CAs: If the trace data includes contract creations by known DApp accounts these newly created CAs *(children)* are added to the set of DApp accounts. b) Parent accounts: If the trace data includes creations of smart contracts that are already known to be part of the DApp, the creating contracts *(parents)* are added to the set of DApp accounts. If any formerly unknown account was added to the set of DApp accounts, the previous three steps are repeated. The goal is to discover the whole create sub-tree of the DApp within the block range. If the input set contains accounts from different create sub-trees, several create sub-trees can be discovered.

That way, several types of information can be retrieved as raw data: 1) The creation sub-tree; meaning all parent or child contracts that belong to the specified set of DApp contracts within the block range. 2) All message calls that were executed in transactions that involved DApp contracts, including those with ETH transfers and function calls with encoded raw data within the block range. 3) Encoded raw data of all log entries that were emitted in a transaction that involved a DApp account in the specified block range.

2) Compute and Transform Execution Data. The raw data of message calls and log entries are then decoded based on the Application Binary Interfaces[10] of DApp CAs that were identified in the transaction traces.

A CA's ABI specifies information to decode data of log entries and function calls in transaction traces. The ABI of a CA holds information about log entries that can be emitted by a smart contract with the exception of log entries according to an ERC-standard and log entries originating from imported libraries in CA code (as of Solidity v0.8.20 all log entries are included in the ABI, see Sect. 6). To maximize the number of retrieved log entries and to expand options for token tracking, ABIs of DApp CAs are expanded by adding standard events of ERC tokens[11]. Additionally, the ABI of a CA holds information about functions that can be called from outside the CA. All in all ABI specifications can be used to decode the log entry and function call raw data using existing libraries[12,13].

---

[9] https://geth.ethereum.org/docs/interacting-with-geth/rpc/ns-debug#debug_tracet ransaction, accessed 2023-04-03.
[10] https://ethereum.org/en/glossary/#abi, accessed 2023-06-03.
[11] https://ethereum.org/en/developers/docs/standards/tokens/, accessed 2023-06-03.
[12] https://github.com/iamdefinitelyahuman/eth-event/blob/master/eth_event/main. py, accessed 2023-03-01.
[13] https://web3py.readthedocs.io/en/latest/web3.contract.html#web3.contract.Contr act.decode_function_input, accessed 2023-03-02.

The extraction method makes use of the third-party service Etherscan[14] which provides access to Ethereum data. We used Etherscan to retrieve the transactions and ABIs of identified DApp CAs.

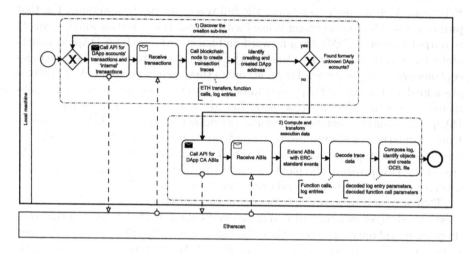

Fig. 4. Extraction method for creating object-centric event logs from blockchain data.

The extracted data is saved tabularly with one row representing a message call or a log entry. We use the PM4Py library function *convert_log_to_ocel()*[15] to convert the tabular data to OCEL. Note that while little domain knowledge is needed to extract the event log data, a certain level of domain knowledge is still needed to format the data to OCEL (e.g., for choosing data attributes as objects).

5 Data Analysis and Evaluation

For an initial validation of the approach, we conduct an assertion [22]. For the assertion, we choose the DApp Augur[16] (v1.0). Augur is an implementation of a betting platform on Ethereum. A bet starts by creating a *market* that contains a statement about a future event. Users can participate in the bet by placing assets in a market. If there is disagreement about the outcome of the event, users can create disputes. Once disputes are resolved, markets finalize and the bet is settled [17]. For Augur, a process mining case study based on a single-case notion event log was presented previously [9]. The previous case study will be used to compare the single notion approach to an object-centric approach in

---

[14] https://etherscan.io/, accessed 2023-06-06.

[15] https://pm4py.fit.fraunhofer.de/static/assets/api/2.7.3/generated/pm4py.convert. convert_log_to_ocel.html, accessed 2023-07-18.

[16] https://augur.net/, accessed 2023-06-04.

terms of data availability and possible insights. To retrieve the object-centric log, we employ the extraction method described in Sect. 4. We will use similar input data as [9]: we extract data starting from Augur's central logging CA[17] and for the blockrange [5926229, 11229573]. The resulting log is available for download[18]. In the following, we describe the resulting data (the creation tree and OCEL event log) and present a brief analysis of the data.

DApp Contracts and Creation Sub-tree. The data extraction was heavily based on identifying CA creations during the observed block range. For Augur, we identified a total of 20866 CAs containing application logic and 1 EOA that served as an initial deployer. The high number of created CAs hints towards extensive use of the factory pattern in Augur's implementation. A visualization of the creation tree (as described in Sect. 3) is depicted in Fig. 5 - nodes represent accounts, and arcs represent creations. The majority of the DApp's CAs were created by a small number of contracts (nodes with many creations are plotted towards the inside of Fig. 5). To better show the tree structure, Fig. 5a contains a subset of 102 creations in which the arcs between creating accounts and created accounts are clearly visible. Fig. 5b shows the entire creation tree. The accounts with the highest number of creations were given designated names on Etherscan. Those include the following accounts with the number of creations in parenthesis: *ShareTokenFactory* (7337)[19], *MapFactory* (3704)[20], *MarketFactory* (2944)[21], *InitialReporterFactory* (2917)[22], *MailboxFactory* (2909)[23], *DisputeCrowdsourcerFactory* (915)[24], *FeeTokenFactory* (142)[25], *FeeWindowFactory* (142)[26], *Augur Deployer* (40)[27], *UniverseFactory* (6)[28], and *ReputationTokenFactory* (2)[29].

Event Log Data. The resulting OCEL-log for Augur comprises 24 activities based on decoded log entries (934167) as well as 1 activity for message calls between accounts (117531). For the block range, all 11 activities of the XES-log from [9] also appear in the OCEL-log. The numbers of the activities' appearance match with one exception: 12 *contribute to dispute* events are missing in the OCEL-log (explanation in Sect. 6). Compared to the XES-log, the OCEL-log has 13 additional activities referring to (a) DApp administration, being *create universe* (1), *create fee window* (116), *sell complete sets* (863), *redeem fee window*

---

[17] 0x75228dce4d82566d93068a8d5d49435216551599.
[18] https://ingo-weber.github.io/dapp-data/augur.html, accessed 2023-06-06.
[19] 0x60a977354a6ba44310b2ee061bcf19632450e51d.
[20] 0x67f53b749fe432274e3f53752a91da89ef86777e.
[21] 0x518530aca60154403012f17c7b8e26f88f7494ee.
[22] 0xbca52c29b535fd63bdc7ca35efa56116550f4c59.
[23] 0xe33ca1ebb783343035b11a7e755c29c28b763540.
[24] 0x1be98680ff697390cbc4cdc414a1be8add733bf7.
[25] 0xe86a4beb10155a5bd7ebb430ce13438341e808a8.
[26] 0x5b4140771615b25f22a4bf52f77e35cdccc5b663.
[27] 0xd82369aaec27c7a749afdb4eb71add9e64154cd6.
[28] 0xe62e470c8fba49aea4e87779d536c5923d01bb95.
[29] 0x8fee0da3a35f612f88fb58d7028d14c7d99a3643.

(a) Subset of 102 creations   (b) 20 866 observed creations

Fig. 5. Creation tree of Augur (blue: DApp CAs; orange: EOA *Augur Deployer* (Color figure online)).

(962), *create order* (24003), *fill order* (15194), and *cancel order* (15150), (b) activities describing token flow, being *burn token* (18456), *give approval* (54934), *mint tokens* (93593), *token was transferred* (284860), and *transfer token* (401854), and (c) activities referring to message and function calls with ETH transfers (117273), of which several are message calls into DApp accounts and which were decoded to retrieve human readable function names and input parameters (39043). The events in the log are described by a total of 95 attributes, from which we created three objects types: *DApp contracts*, *DApp users* and *markets*. We reckon that other object types including *transaction* or *block* might be beneficial for specialized types of analysis, too (e.g., for security analysis).

Analysis. For the analysis, we filtered the OCEL-log to include all common events with the XES-log, additional log entries from Augur's central logging CA, and token transfers. We included objects for *markets* and *DApp accounts*. The resulting object-centric directly-follows graph is depicted in Fig. 6. In addition to the results from [9] we can see that events occur that are not related to a *market* but are triggered by a DApp contract, e.g., *create universe*, *create fee window*, and *create order*. The event *transfer token* is only rarely triggered by the main logging contract. Instead a high number of other DApp accounts manage the token distribution in Augur.

As described in Sect. 2, a central advantage of object-centric logging formats is avoiding deficiency, convergence, and divergence that can happen when flattening data to suit a single case notion. We investigated if these advantages apply to the OCEL-log we extracted. In terms of *deficiency* we found activities that relate only to a selection of object types. E.g., the *market* notion does not cover the activities *create universe*, or *create fee window*. Additionally, token transfers between DApp accounts and user accounts cannot be related to markets either. The same is true for Ether transfers between accounts. In particular, when transfers and Gas fees are involved, deficiency should be avoided

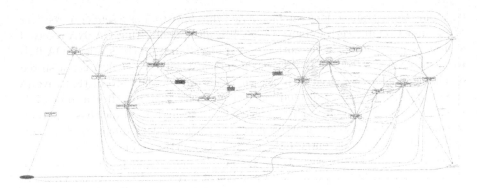

Fig. 6. Object-centric directly-follows graph discovered with contracts, markets, and transactions as objects.

to cover process costs accurately. The Augur process had a total transfer volume of 349131 ETH between accounts during DApp execution (note that during one transaction ETH can be transferred from account to account and might count several times) which can go unnoticed focusing on the market case notion only. *Convergence* would appear in the Augur log mainly for activities relating to the market as well as the user account notion. Those include, e.g., *create market*, *redeem as initial reporter*, *create dispute*, etc., all of which would have to be duplicated to represent the process in two event logs. *Divergence* becomes an issue depending on the chosen notion. Events could appear related through a common CA object although they are not. To illustrate the divergence issue with Augur as an example: Augur is one DApp consisting of multiple CAs. The code in the CAs execute different parts of the application. E.g., market behavior is processed in CA 0xA. In parallel an incentive system with a native token is implemented in CA 0xB. If the single case notion *DApp account* was selected, different token transfers between different users that were active in different markets might appear related, although they are not.

6 Discussion

The analysis in the previous section illustrated benefits of using an object-centric logging format over a single case notion format. Applying our approach and the subsequent analysis also uncovered several points to debate.

The current implementation of this paper's data extraction method relies on the third-party service Etherscan. The source code of Etherscan is not fully public so its use for data retrieval reduces transparency in the research process. However, that dependence was accepted for two reasons: 1) *Etherscan is an indexed data store of the Ethereum blockchain*. To reduce the size of the Ethereum blockchain, only the results of transactions are stored on-chain, while the operations leading up to the transactions' outcomes (transaction traces) are not. For querying data from a blockchain archival node that means: for a CA

0xA, only those transactions can be queried that have 0xA as a sender or receiver of a transaction in a certain block range. It is also possible that CA 0xA received a message call during the execution of a transaction from the sender EOA 0xB to the receiver CA 0xC. In that case, the message call 0xC $\rightarrow$ 0xA is not stored on-chain. Indexing services such as Etherscan provide data stores that save (a subset of) such message calls and make them easily queriable. It is also possible to circumvent using a third-party service by creating a local database containing message calls between all active accounts within a block range. 2) *Retrieving DApp CA's ABIs.* Log entry data as well as function names and inputs are stored on the blockchain encoded. Decoding the data requires information from the CA's ABI. Etherscan provides a number of verified ABIs for CAs, which we drew from. If the smart contract code of the CA is known, the ABI can also be computed by a Solidity compiler.

To discover all accounts belonging to a DApp, the extraction method hinges on knowing at least one contract of each creation sub-tree. If multiple root deployers exist, creation sub-trees may partially not be discovered. Note, however, that transaction traces of undiscovered sub-tree accounts can still be logged, if they contain message calls or creations by known DApp accounts. That was also the case in the Augur log. 12 *contribute to dispute* events were missing in our extracted data. We investigated the issue and found a second deployer EOA that is not labeled on Etherscan[30]. As a result, a fraction of the DApps transactions could not be discovered. Vice versa, if a root deployer account was used to deploy more than a single DApp, the extraction method could discover accounts that do not belong to the DApp and mistakenly extract non-DApp-account data. Precautions can be taken by choosing distinguished block ranges during extraction or explicitly excluding accounts from considered candidates of DApp accounts. Both countermeasures require a level of domain knowledge though.

Furthermore, the extraction method assumes transparency about log entries of a CA. Types of log entries are documented in a CA's ABI. In Solidity, ABIs only include information about log entries that are explicitly included in a CA's smart contract code and inherited log entries of another smart contract[31]. There may, however, be log entries emitted by library code of a CA, which were not included in the ABI. This paper's extraction method thus, could not capture library log entries for Augur. Library log entries were only recently included in the ABI for new Solidity compiler versions[32] (as of Solidity v0.8.20[33]). For smart contracts written in older Solidity code, however, the issue persists.

This paper's approach was implemented with the Geth-based Ethereum client Erigon, so that the rich features of Geth were available for transaction trace generation. We cannot make a statement about reproducibility with other clients except Geth and Erigon. However, both clients have a combined market share

---

[30] 0x57f1c2953630056aaadb9bfbda05369e6af7872b.
[31] https://github.com/ethereum/solidity/issues/13086, accessed 2023-06-03.
[32] https://github.com/ethereum/solidity/pull/10996, accessed 2023-06-03.
[33] https://github.com/ethereum/solidity/releases/tag/v0.8.20, accessed 2023-06-06.

of ca. 70% of all Ethereum clients, which makes the approach accessible for the majority of node operators[34].

We also noticed limitations when using OCEL. In blockchain applications such as Augur, different types of objects exist, e.g., DApp CAs, transactions, user accounts, etc. Some of these objects might change their role as a process instance progresses. E.g., the same user account in Augur can be sender as well as receiver of a token. Documenting an object's changing roles over time is currently not explicitly supported in OCEL. Hence an integrated picture showing full traces of an object with its changing roles cannot be depicted. The same phenomenon was described by [12,15] as an issue in representing "object evolution".

Another challenge for log generation is determining the order of events in a blockchain environment. In past studies, a blockchain event's timestamp was defined as the timestamp of the block it was included in. There may, however, be an order to events or transactions within a block. Similarly, EVM traces of individual transactions represent a (tree) structure with an order of events. Events and message calls in the same block appear to have occurred at the same time when only considering the inclusion block's timestamp. OCEL requires a timestamp to determine the order of events and does not allow additional complementary ordinal variables. We hence had to manipulate the timestamps to preserve the events' and message calls' order.

The presentation of the paper's approach is heavily based on the technology of Ethereum blockchains. In addition to the Ethereum Mainnet, from which we extracted the data, the approach is also applicable on other Ethereum-based networks used in enterprises. Some of the concepts we made use of are transferable to other blockchains. For one, executable code is deployed as smart contracts also in other second generation blockchains (e.g., Hyperledger Fabric[35] and can be grouped to identify code belonging to one application. Also the notion of transactions of assets is a concept shared by all blockchains [21, p.5] and can be exploited to retrieve object-centric process mining logs across platforms.

7   Conclusion

Based on the literature, we showed shortcomings of the single notion logging format XES to capture execution data of DApps. We presented an approach to retrieving execution data from dynamically deployed blockchain applications with little prior knowledge about the application. Based on one case application, we examined the suitability of the current object-centric logging standard in process mining as an event log format for blockchain applications' execution data. For the considered case, we observe a first indication that OCEL is a suitable format with respect to mapping the distributed nature of blockchain logging in different accounts and varying levels of object types (CAs, EOCs, tokens, transactions, etc.) and tackles deficiency, convergence, and divergence issues.

---

[34] https://clientdiversity.org/#distribution, accessed 2023-06-03.

[35] https://hyperledger-fabric.readthedocs.io/en/latest/smartcontract/smartcontract. html, accessed 2023-07-18.

We also note that OCEL appears to be unsuitable for depicting the transition of objects between roles within the same activity, e.g., an EOC being a sender in one event *Transfer* and a receiver in another event *Transfer*.

References

1. van der Aalst, W.M.P.: Process mining: a 360 degree overview. In: van der Aalst, W.M.P., Carmona, J. (eds.) Process Mining Handbook. LNBIP, vol. 448, pp. 3–34. Springer, Cham (2022). https://doi.org/10.1007/978-3-031-08848-3_1
2. Van der Aalst, W.M., et al.: IEEE standard for extensible event stream (xes) for achieving interoperability in event logs and event streams. IEEE Std. **1849–2016**, 1–50 (2016)
3. Aalst, W.M.P.: Object-centric process mining: dealing with divergence and convergence in event data. In: Ölveczky, P.C., Salaün, G. (eds.) SEFM 2019. LNCS, vol. 11724, pp. 3–25. Springer, Cham (2019). https://doi.org/10.1007/978-3-030-30446-1_1
4. Bandara, H.D., et al.: Event logs of Ethereum-based applications (2021)
5. Beck, P., et al.: BLF: a blockchain logging framework for mining blockchain data. In: BPM (PhD/Demos), pp. 111–115 (2021)
6. Corradini, F., Marcantoni, F., Morichetta, A., Polini, A., Re, B., Sampaolo, M.: Enabling auditing of smart contracts through process mining. In: ter Beek, M.H., Fantechi, A., Semini, L. (eds.) From Software Engineering to Formal Methods and Tools, and Back. LNCS, vol. 11865, pp. 467–480. Springer, Cham (2019). https://doi.org/10.1007/978-3-030-30985-5_27
7. Ghahfarokhi, A.F., Park, G., Berti, A., van der Aalst, W.M.P.: OCEL: a standard for object-centric event logs. In: Bellatreche, L., et al. (eds.) ADBIS 2021. CCIS, vol. 1450, pp. 169–175. Springer, Cham (2021). https://doi.org/10.1007/978-3-030-85082-1_16
8. Günther, C.: XES standard definition. http://www.xes-standard.org/ (2009)
9. Hobeck, R., Klinkmüller, C., Bandara, H.D., Weber, I., van der Aalst, W.: Process mining on blockchain data: a case study of Augur. In: BPM2021: International Conference on Business Process Management, pp. 306–323. Rome, Italy (2021)
10. van der Aalst, W., et al.: Process mining manifesto. In: Daniel, F., Barkaoui, K., Dustdar, S. (eds.) BPM 2011. LNBIP, vol. 99, pp. 169–194. Springer, Heidelberg (2011). https://doi.org/10.1007/978-3-642-28108-2_19
11. Klinkmüller, C., Ponomarev, A., Tran, A.B., Weber, I., van der Aalst, W.M.P.: Mining blockchain processes: extracting process mining data from blockchain applications. In: BPM Blockchain Forum, pp. 71–86 (2019)
12. M'Baba, L.M., Assy, N., Sellami, M., Gaaloul, W., Nanne, M.F.: Extracting artifact-centric event logs from blockchain applications. In: 2022 IEEE International Conference on Services Computing (SCC), pp. 274–283. IEEE (2022)
13. M'Baba, L.M., Sellami, M., Gaaloul, W., Nanne, M.F.: Blockchain logging for process mining: a systematic review. In: HICSS 2022: 55th Hawaii International Conference on System Sciences, pp. 6197–6206. HICSS (2022)
14. Mendling, J., et al.: Blockchains for business process management: challenges and opportunities. ACM Trans. Manag. Inf. Syst. **9**(1), 1–16 (2018)
15. Moctar M'Baba, L., Assy, N., Sellami, M., Gaaloul, W., Farouk Nanne, M.: Process mining for artifact-centric blockchain applications. Simul. Model. Pract. Theory **127**, 102779 (2023). https://doi.org/10.1016/j.simpat.2023.102779. https://www.sciencedirect.com/science/article/pii/S1569190X23000564

16. Mühlberger, R., Bachhofner, S., Di Ciccio, C., García-Bañuelos, L., López-Pintado, O.: Extracting event logs for process mining from data stored on the blockchain. In: Business Process Management Workshops, pp. 690–703 (2019)
17. Peterson, J., Krug, J., Zoltu, M., Williams, A.K., Alexander, S.: Augur: a decentralized oracle and prediction market platform. Tech. rep., Forecast Foundation (2018). https://github.com/AugurProject/whitepaper/blob/master/v1/english/whitepaper.pdf. Accessed 05 Jan 2021
18. Wirawan, N.Y., Yahya, B.N., Bae, H.: Incorporating transaction lifecycle information in blockchain process discovery. In: Lee, S.-W., Singh, I., Mohammadian, M. (eds.) Blockchain Technology for IoT Applications. BT, pp. 155–172. Springer, Singapore (2021). https://doi.org/10.1007/978-981-33-4122-7_8
19. Wood, G., et al.: Ethereum: a secure decentralised generalised transaction ledger. Ethereum Proj. Yellow Paper **151**(2014), 1–32 (2014)
20. Xu, X., Pautasso, C., Zhu, L., Lu, Q., Weber, I.: A pattern collection for blockchain-based applications. In: Proceedings of the 23rd European Conference on Pattern Languages of Programs. EuroPLoP 2018, Association for Computing Machinery, New York, NY, USA (2018). https://doi.org/10.1145/3282308.3282312. https://doi.org/10.1145/3282308.3282312
21. Xu, X., Weber, I., Staples, M.: Architecture for blockchain applications, 1st Edn. Springer, Cham (2019). https://doi.org/10.1007/978-3-030-03035-3
22. Zelkowitz, M., Wallace, D.: Experimental models for validating technology. Computer **31**(5), 23–31 (1998). https://doi.org/10.1109/2.675630

RPA Forum

Preface

Robotic Process Automation (RPA) is a maturing technology in the field of Business Process Management. Plattfaut & Borghoff (2022) defined it as "a technology that allows the development of (multiple) computer programs (i.e., bots) that automate rules-based business processes through the use of GUIs". As such, it enables the office automation of repetitive tasks.

The technology itself has reached a certain level of technological maturity and organizational adoption. This means that researchers now have the chance to look at RPA in a larger context. Examples include

- Low-code automation: RPA is part of a development towards low-code automation aimed at building and automating processes with off-the-shelf software solutions that do not require extensive programming skills.
- RPA for smart automation: RPA can be combined with other technologies, such as process mining, AI, ML, eye-tracking, OCR, or chatbots, with the goal of a more flexible, holistic process automation.
- Organizational implications and considerations: RPA leads to tremendous advantages, but business cases and effects on organizational knowledge and capabilities are unclear.

Topics like these are studied by a broad range of researchers across communities. While computer scientists might focus on the technological aspects, management researchers study the impact of RPA on organizations. Information systems scholars might focus on the socio-technological nature.

The Robotic Process Automation Forum of the 21st International Conference on Business Process Management aims to provide a forum for this broad range of researchers to exchange ideas, fuel new research, and start corresponding collaborations. To this end, a panel discussion on the future of RPA research accompanied the paper presentations. In total, the forum attracted 21 international submissions on a diverse range of topics around RPA. All submissions were reviewed by three Program Committee members which led to an acceptance of eight papers. These eight papers show the full range of topics mentioned above.

The first paper by Mirispelakotuwa et al. addresses the emerging organizational challenge of process knowledge loss (PKL) that can occur with the increasing adoption of RPA. Based on a literature review and expert interviews, the authors develop an empirically supported conceptual model highlighting nine factors that impact PKL in the context of RPA, including positive factors (e.g., top management support), negative factors (e.g., employee turnover), and factors with both positive and negative impacts (e.g., RPA governance). Their model can provide valuable insights and strategies to address PKL issues associated with RPA implementation.

In their paper, Helbig & Braun focus on the business models for RPA, i.e., the strategies that companies can employ to make RPA financially attractive. Also based on a literature review and expert interviews, the authors propose an approach to establish

RPA as a driver of digitization and automation within a company, using the Business Model Canvas as an analysis tool for a holistic and iterative view of business models. Evaluated in a case study with an internal IT service provider, their approach may assist practitioners in gaining a multi-perspective view of RPA implementations.

Průcha & Madzík address an organizational issue in RPA governance, namely the reusability of RPA components, by proposing SiDiTeR (Similarity Discovering Techniques for RPA). These techniques identify similarities between RPA processes using source code or process logs. In their evaluation on publicly available process designs, they found 655 matches across 156 processes. These results showcase the significance of process similarity in mitigating the maintenance burden associated with RPA.

Motivated by the challenge of extracting meaningful insights from complex and information-dense graphical user interfaces (UIs), Martíinez-Rojas et al. introduce a novel approach that utilizes eye-tracking technology for RPA. Their proposed solution incorporates gaze fixation data into UI logs, allowing for the filtering of irrelevant information. In a preliminary evaluation, the method successfully reduces irrelevant UI elements by an average of 76%, while retaining meaningful information on the screen, stressing the potential of eye-tracking data as another relevant data source for RPA.

A novel application case for RPA stands in the center of the paper by Elsayed et al., which explores the use of Robotic Process Automation (RPA) as a solution to partially automate Computerized System Validation (CSV). CSV is required to adhere to regulatory requirements in many industries, but its enormous costs pose a challenge for many companies. The ongoing research project applies RPA to partially automate CSV, effectively reducing time and effort associated with manual CSV activities and showcasing the potentials of RPA beyond the typical application cases.

In their paper, Strothmann & Schulte examine the transition from frontend automation RPA to backend automation with APIs, considering the temporary nature of RPA as a technology gap-filler. Insights from interviews and literature reviews reveal the requirements for designing migration-ready RPA bots, challenges to overcome, and selection criteria for prioritization. The findings are integrated into BPM methodologically and architecturally, and a prototypical implementation in the telecommunications industry demonstrates the migration from frontend to backend automation in an organizational setting.

The objective of Huo et al. is to facilitate the use of the APIs of conversational interfaces, such as chatbots, by automation tools, such as unattended RPA bots, enabling organizations to leverage enterprise tools and automations more effectively. To achieve this, they propose a data augmentation approach based on paraphrasing using large language models and a system to generate sentences to train an intent recognition model. Experimental results demonstrate the effectiveness of the proposed approach, which has been deployed in a real-world setting, enhancing user experience and satisfaction.

Finally, Wiethölter et al. present a design science research project on "customer journey mining," an approach that leverages digital customer data for predicting customer movements through individual journeys. They operationalize the developed customer journey mining artifact through a case study of an online travel agency, training seven prediction models, with gradient-boosted trees yielding the highest accuracy at 43.1%.

The research also explores the potential interplay of RPA and customer journey mining, finding technical suitability for RPA implementation but limited financial viability.

We would like to thank all authors who submitted their work to the RPA forum. Moreover, we also express our gratitude to the program committee: Simone Agostinelli, Aleksandre Asatiani, Bernhard Axmann, Adela del Río-Ortega, Carmelo Del Valle, Mathias Eggert, Carsten Feldmann, Peter Fettke, Christian Flechsig, Norbert Frick, José González Enríquez, Lukas-Valentin Herm, Hannu Jaakkola, Christian Janiesch, Andrés Jiménez Ramírez, Fabrizio Maria Maggi, Andrea Marrella, Massimo Mecella, Dan O'Leary, Teijo Peltoniemi, Yara Rizk, Rehan Syed, Maximilian Völker, and Moe Wynn. Last, we thank the BPM 2023 chairs and organizers for their work on the overall conference and generous chance to host the RPA forum.

Organization

Program Chairs

Ralf Plattfaut — University of Duisburg-Essen, Germany
Jana-Rebecca Rehse — University of Mannheim, Germany

Program Committee

Simone Agostinelli	Sapienza University of Rome, Italy
Aleksandre Asatiani	University of Gothenburg, Sweden
Bernhard Axmann	Technical University of Ingolstadt, Germany
Adela del Río-Ortega	University of Seville, Spain
Carmelo Del Valle	University of Seville, Spain
Mathias Eggert	FH Aachen, Germany
Carsten Feldmann	FH Münster University of Applied Sciences, Germany
Peter Fettke	German Research Center for Artificial Intelligence (DFKI), and Saarland University, Germany
Christian Flechsig	Technische Universität Dresden, Germany
Norbert Frick	Hochschule der Deutschen Bundesbank, Germany
José González Enríquez	University of Seville, Spain
Lukas-Valentin Herm	Julius-Maximilians-Universität Würzburg, Germany
Hannu Jaakkola	University of Tampere, Finland
Christian Janiesch	TU Dortmund University, Germany
Andrés Jiménez Ramírez	University of Seville, Spain
Fabrizio Maria Maggi	Free University of Bozen-Bolzano, Italy
Andrea Marrella	Sapienza University of Rome, Italy
Massimo Mecella	Sapienza University of Rome, Italy
Dan O'Leary	University of Southern California, USA
Teijo Peltoniemi	University of Turku, Finland
Yara Rizk	IBM Research, USA
Rehan Syed	Queensland University of Technology, Australia
Maximilian Völker	Hasso Plattner Institute, Germany
Moe Wynn	Queensland University of Technology, Australia

Is RPA Causing Process Knowledge Loss? Insights from RPA Experts

Ishadi Mirispelakotuwa[✉][iD], Rehan Syed[iD], and Moe T. Wynn[iD]

Queensland University of Technology, Brisbane, Australia
ishadi.mirispelakotuwa@hdr.qut.edu.au, {r.syed,m.wynn}@qut.edu.au

Abstract. Robotic process automation (RPA) is a process automation technology that mimics human behaviour using software agents called 'bots'. This study aims to explore one of the emerging organisational challenges of RPA use—process knowledge loss (PKL), which materialises when bots start performing repetitive and rule-based tasks, replacing employees. PKL can negatively impact an organisation's continuous improvement of processes, productivity, and competitiveness. Thus, it is a critical area that requires scholarly attention. There is a dearth of studies focusing on knowledge loss issues in RPA. Hence, no empirical models or frameworks exist to explain RPA's impact on PKL. To address this research gap, we first reviewed RPA literature. Then, the findings were further investigated using seven RPA expert interviews. The RPA experts confirmed the existence of PKL in the context of RPA and explained the influencing factors. We present an empirically supported conceptual model illustrating how RPA impacts PKL, which goes beyond highlighting the phenomenon as a process-related or knowledge-management challenge. The conceptual model captures ten factors, including three positive factors that mitigate PKL (i.e., top management support, process expertise, and RPA-BPM integration), four negative factors that contribute to PKL (i.e., employee turnover, knowledge hiding, automation complacency, and continuous process redesign), and three factors with both positive and negative impacts (i.e., employee redeployment, RPA governance, and task division). These findings contribute to the knowledge base on RPA associated with PKL. This model may assist organisations in devising strategies to mitigate RPA-related PKL.

Keywords: Robotic Process Automation · Process Knowledge Loss · Process Knowledge · Expert Interviews · Conceptual Model

1 Introduction

Organisations are increasingly leveraging business process automation (BPA) to transform and enhance organisational processes [12]. Robotic Process Automation (RPA) is a task-level, low-code automation technology that uses 'bots' (a.k.a. software robots) to emulate manual, repetitive, and rule-based tasks through graphical user interfaces [42]. RPA research to date has mostly focused

J. Köpke et al. (Eds.): BPM 2023, LNBIP 491, pp. 73–88, 2023.
https://doi.org/10.1007/978-3-031-43433-4_5

on how organisations and their employees can benefit from RPA deployments. For example, studies show that RPA increases the operational efficiency and traceability of organisational processes, ensures business continuity, and increases job satisfaction among employees [4, 9, 18, 24, 27].

Despite these benefits, recent research and practice commentary point to various challenges and negative consequences of RPA [26, 30]. Process knowledge loss (PKL) is one such negative consequence of RPA highlighted by recent literature [5, 6, 14, 26, 30, 45] and RPA practitioners [16]. According to Marciniak and Stanisławski [26], an organisation may experience process "amnesia" over time if a task is entirely automated with RPA and no longer performed manually. Similarly, Eulerich et al. [14] stated that PKL emerges when a bot executes an entire process, resulting in the organisation losing the ability to carry out these tasks without the support of bots. Both definitions assume that an end-to-end automation of a process without any human touch-points leads to PKL. However, practice literature [16] stated that most organisations follow a hybrid model where process execution is shared between employees and bots. Thus, the concept of PKL requires further clarification. Considering the above notion, this study defines RPA-related PKL as follows. *RPA-related PKL is the intentional or unintentional loss of knowledge related to the process resulting from using RPA, where bots start performing repetitive, rule-based tasks previously handled by employees.* The rationale behind the definition is explained as follows. First, PKL is identified as a subset of organisational knowledge loss (OKL). OKL refers to the loss of internally established knowledge intentionally or unintentionally [10, 21]. Intentional knowledge loss is an organisational attempt to intentionally forget knowledge unfavourable to organisational performance. Unintentional knowledge loss refers to the accidental loss of knowledge [21]. An established form of existing organisational knowledge can be embedded in durable organisational objects such as culture, values, processes, databases, etc. [21]. Hence, it is possible to describe a great deal of organisational knowledge (more than 90% in most cases) in terms of processes [3]. Accordingly, process knowledge can be identified as a branch of organisational knowledge [3, 11]. Following this line of reasoning, PKL can be defined as *the intentional or unintentional loss of knowledge related to processes.* Next, we aligned the definition of PKL to the RPA context. Our definition addresses the limitation of [26]'s and [14]'s explanation of PKL by considering the broad use of RPA in a hybrid model and end-to-end automation.

The study also differentiates PKL from the notion of deskilling [5]. Deskilling refers to reducing or eliminating skilled labour due to increased automation [1]. Asatiani et al. [4] stated that process automation may lead to deskilling as employees no longer need the skills to perform a particular task. Accordingly, the central focus of deskilling is on the loss of skills at an individual level, whereas PKL centers around the organisational-level loss of process knowledge.

PKL can result in several negative organisational impacts. One impact is that it impedes continuous improvement. Without process expertise, organisations struggle to identify process improvement opportunities. Employees' lack of process knowledge can hinder continuous improvement initiatives and limit the organisation's ability to adapt and innovate [17,37]. Decreasing organisational productivity is also a concern. RPA bots may encounter exceptions that were not explicitly covered during development [20,35]. Without sufficient process knowledge among employees, the organisation may experience difficulties in troubleshooting errors and handling exceptions [33]. This can lead to process disruptions, decreasing the process performance and negatively impacting the customer experience [30]. Additionally, an organisation may incur costs and time investment in training employees to mitigate this issues [26]. Thus, PKL remains a critical concern in the context of RPA that needs to be addressed.

To date, the research on RPA-related PKL has been limited to anecdotal evidence from industry reports such as [16] and a few empirical studies that highlight selected facets of the phenomenon [5,6,14,26,30,45]. The existing studies also have limited evidence with a high-level observation describing PKL in the context of RPA as a process-related [5,6,30] or knowledge management challenge [14,26,45]. Hence, there are no empirical models or frameworks that systematically capture the potential impact of RPA-related PKL. However, PKL is a phenomenon that can have significant negative consequences on organisations [17,37]. Thus, it necessitates the need for empirical validation to strengthen the understanding of the phenomenon in order to optimise the use of RPA as a technology investment. Therefore, this study sets out to answer the following research question: *how does RPA impact PKL in organisations?*

The study presents an empirically validated conceptual model. First, an initial conceptual understanding was developed using RPA literature. Next, primary data was collected from seven RPA experts. Then, a conceptual model was developed with ten factors impacting the RPA-related PKL. This study contributes to the RPA knowledge base by exploring the impact of RPA on PKL, which can also be generalised to other BPA initiatives. This model will help organisations strategise to tackle PKL-related challenges.

The rest of the paper unfolds as follows. Section 2 summarises the study design. Section 3 presents related work within RPA literature. Section 4 provides a synthesis of interview findings. Section 5 presents a discussion of the developed conceptual model with the limitations of the study. Section 6 concludes the paper with suggestions for future work.

2   Study Design

The study design is comprised of 2 stages. Firstly, a systematic literature review [7] was conducted to build the foundational knowledge. Figure 1 depicts the number of articles that resulted in the literature search process. The queries used for the literature search incorporated the following keywords and synonyms, namely, robotic process automation (synonyms included desktop automation, low code automation, and software robots), knowledge loss (or knowledge management), and process knowledge. In total, 51 articles were analysed and inductively coded following the guidelines of [36]. The analysis resulted in a total of nine themes from the literature, which is briefly discussed in Sect. 3.

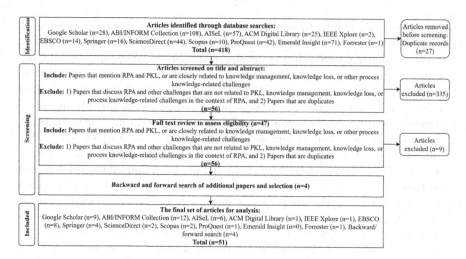

Fig. 1. Literature Search Process

Literature findings were further investigated using seven RPA expert interviews.[1] We employed purposive sampling to select experts from different sectors, as shown in Table 1. A hybrid approach combining an inductive and deductive approach was used to analyse primary data [36,44]. First, nine themes discovered using literature were used to deductively analyse and classify open codes generated via primary data. All themes that emerged from the literature were confirmed using primary data with additional insights. Next, coding was performed inductively by grouping the new open codes. An additional theme - continuous process redesign emerged at this stage. By combining the findings of both primary and secondary data, a conceptual model was developed with ten positive/negative themes as depicted in Fig. 2 and explained in detail in Sect. 5. During code extraction, a coding rule book was developed to ensure a formalised approach [36]. All coding rounds relied on coder corroboration with a second coder and a third coder using a critical review process.

3 Literature Review Findings

Although there are no studies specifically examining how RPA impacts PKL, several studies have discussed various aspects of PKL as process-related or knowledge-management challenges caused by RPA. This has paved the way for reviewing RPA literature. Accordingly, nine themes emerged from the literature, namely, 1) employee turnover, 2) knowledge hiding, 3) automation complacency, 4) top management support, 5) employee redeployment, 6) RPA governance, 7) task division, 8) process expertise, and 9) RPA-BPM integration. All related references for each theme are provided in the supplementary material - Part B.[2]

---

[1] Supplementary Materials - Part A - Interviewee Profiles and Details.
[2] Supplementary Materials - Part B - Evidence of Themes from Literature.

Table 1. Interviewee Details.

Role	Sector	Duration	Experience	Code
Automation Architect	Healthcare	50 min	6 years	I1
Principal RPA Developer	Education	55 min	5+ years	I2
Senior RPA Developer	Healthcare	45 min	4 years	I3
Business Improvement Consultant	Local Government	35 min	7+ years	I4
Head of Automation	IT Consulting	45 min	7 years	I5
Project Lead	Insurance	50 min	5+ years	I6
Project Manager	Insurance	50 min	3+ years	I7

Employee turnover emerged as a theme defining RPA-related PKL. RPA was considered to cause employee layoffs due to the replacement of software bots with human workers [13,15,30]. Employee turnover was discussed as a general cause of knowledge loss within organisations, which can also be seen in the RPA context. Typically, departing employees carry subject-matter expertise. This also disrupts social networks, ultimately impacting organisational knowledge exchange [14,16,38,39]. Accordingly, the existing literature demonstrates that RPA causes employee turnover, which results in PKL.

Knowledge hiding or employee resistance to sharing knowledge was identified as a driver of PKL. Intra-organisational knowledge flows are hindered by knowledge hiding due to poor knowledge-sharing practices [38,39]. Job insecurity is a major concern that can trigger resistance to knowledge sharing in RPA. Employees believe that having process expertise will prevent them from being replaced by bots [25,28,30]. As a result, knowledge hiding can be identified as a factor impacting PKL in the context of RPA.

Automation complacency occurs when human agents are less likely to exert supervisory control as a result of excessive reliance on automation [5]. Several studies noted that automation complacency causes employees to lose their fundamental understanding of the process logic and their hands-on end-to-end understanding of the process [5,30]. Despite a brief discussion in the literature, there is a clear link between automation complacency and RPA-related PKL.

Top management support was identified as a critical success factor (CSF) in RPA [33], but little was discussed on how the involvement of senior managers impacts knowledge management in RPA initiatives [5,28,30,33,38]. Proactive engagement of senior management in skill development and continuous knowledge management was acknowledged as reducing knowledge retention issues within RPA initiatives [28,30,33]. However, no study has specifically investigated how top management's support could impact the RPA-related PKL. Thus, the field warrants further empirical studies to support and clarify the existing link between top management support and PKL in the context of RPA.

Employee redeployment emerged as a theme that refers to changing job profiles and reemploying employees within the organisation or in the same team. Studies highlighted that employees' job profiles changed when they were replaced by bots [4,13,15,27,30,31,46]. RPA enables employee upskilling and reorienta-tion into more creative and value-adding roles, which include supervising bots and maintaining task control [4,15,27,30]. Despite the absence of directed stud-ies that examine RPA's real impact on PKL due to employee redeployment, existing literature suggests that it may have a positive impact on RPA-related PKL. However, further empirical data is required to validate the emerging link between employee redeployment and RPA-related PKL.

RPA governance was identified as a mechanism to overcome the issue of RPA-related PKL [16,26,42]. Ensuring RPA governance through audit trials and detailed process documentation was suggested as a solution to overcome the issue of PKL [14,16,26,42]. A general discussion of RPA governance estab-lished through a centre of excellence (CoE) was found in the literature pertain-ing to knowledge management of RPA projects [5,15,19,20,29,30,32]. There are three types of RPA governance models: centralised, decentralised, and feder-ated [22,29]. In a centralised governance approach, knowledge is centralised in knowledge repositories in terms of process designs, automation rulebooks, and algorithms [5,19]. Due to this, centralised governance was favourably viewed toward knowledge management in RPA [26]. In contrast, studies also revealed that the centralisation process of governance hinders collaboration between local units and the central hub and does not immediately attend to knowledge require-ments [5]. Accordingly, the literature views centralised governance both posi-tively and negatively. According to the literature, decentralised RPA governance negatively impacts knowledge management since capabilities are spread across departments [29]. It was found that federated governance was more effective in maintaining and disseminating knowledge among RPA projects as it could benefit from both centralised and decentralised structures [29]. Therefore, it was identified as having a positive impact on knowledge dissemination in RPA. However, the literature lacks evidence to support how each governance model specifically impacts PKL.

Task division refers to assigning appropriate tasks to employees and bots [2,4–6,8,13,14,16,25,26,28,30,35]. The most common way of dividing tasks among employees and bots is based on mindfulness, where mindful tasks (that require creativity) are assigned to employees and mindless tasks (rou-tine tasks that do not depend on human cognition) are assigned to bots [5]. As RPA is equally capable of retaining implicit process knowledge in the form of workflow specifications, a task division among employees and bots appears to be favourable for retaining process knowledge [8,13,34,38–40]. Several stud-ies [5,6,14,26,30,31] and industry insights [16] counter-argue that task division leads to PKL, specifically due to task visibility issues that result from black box nature of tasks or fragment of the process executed by bots [5,6,30,43]. There-fore, there is no consensus regarding task division and PKL in the literature.

Process expertise must be preserved for maintaining and transferring crit-ical process knowledge [5,18,19,30,33,42,46]. Plattfuat et al., [33] emphasised expert support and process knowledge as two CSFs in RPA. Typically, when employees with such process expertise are replaced by bots, they are redeployed in different roles as 'process champions' or 'process leaders' to retain their knowledge within the organisation [30,32]. The support of these process experts then becomes necessary when training new employees or disseminating knowledge in RPA projects [19,31,42]. As a result, process expertise is identified as having a positive impact on mitigating knowledge management issues in RPA.

RPA-BPM integration is an emerging discussion [19,23,24,35,41]. RPA does not replace BPM, but rather complements it [23]. Several studies proposed that, as a more established research field, BPM has the potential to provide an environment for technologies like RPA to thrive [19,23]. BPM synergises knowledge management and processes [41]. Thus, when RPA is integrated into BPM, the limitations of RPA can be overcome, and the process knowledge needed for successful RPA realisation can be provided [23]. It has been briefly discussed how RPA-BPM integration can mitigate knowledge management issues, but the evidence indicates that there is a weak link between RPA-BPM integration and PKL, requiring further research.

In summary, several themes have clear evidence that points to their impact on PKL, while others have limited evidence requiring further empirical validation.

4   Interview Findings

The following section summarises the empirical findings from the seven RPA expert interviews. The analysis supported validating all nine themes identified in the literature while revealing an additional theme - continuous process redesign. Each theme is supported by relevant evidence from primary data.

Due to **employee turnover**, organisations face difficulties maintaining the process due to a loss of process-related knowledge. *"Once that person [process expert] leaves, no one will actually be able to maintain that process"* (I5). A participant highlighted that process expertise could be lost due to employee turnover. *"If a particular resource is, moving or moving out of the organisation, resigning, or moving to a totally different job role, then that person is taking away the entire knowledge of how the rules, how the process was designed, and how the bot is working, and that entire knowledge is removed from that environment"* (I7). Accordingly, employee turnover was discussed as a common phenomenon that occurs independent of RPA adoption but can negatively impact PKL.

Knowledge hiding refers to the unwillingness of an employee to share their knowledge acquired through experience. A participant highlighted this concern, stating that employees hide their knowledge due to job insecurity concerns caused by technologies that mimic human behaviour, such as RPA. *"The previous team may not necessarily disclose the information about the process as necessary, because maybe, you know, he wants to protect the job or whatever"* (I5). Findings reveal that RPA adoption can trigger employee knowledge-hiding behaviour, which can eventually result in PKL.

Automation complacency was manifested among employees as a state of feeling relaxed and not having to pay much attention to or take responsibility for the tasks performed by bots. *"Due to that [relying on RPA], the major* *knowledge-related issue happens...I saw a decline in responsibility because they have it in their mind that the bot is there and the bot is doing the job"* (I7P). Employees who become complacent lose their process knowledge due to a lack of hands-on execution, resulting in an inability to recall steps. *"If you're not doing the process and you don't talk about that process anymore, letting the bot do everything...within six months or a span of one year of time, you tend to forget how to do that [process]"* (I1). Accordingly, high reliance on RPA can result in PKL.

Top management support for reducing PKL in the context of RPA was highlighted by several participants. Top management support includes the proactive engagement of senior managers for knowledge management across RPA initiatives in terms of setting up documentation standards, conducting knowledge audits, and providing necessary training for RPA stakeholders. *"When we hand over the bot for the first time when the process is going into production, we were asked [by senior managers] to give a training or workshop to the business stakeholders"* (I1). Accordingly, top management support was conferred as a factor that mitigates knowledge management challenges associated with RPA.

Employee redeployment refers to the concept of reassigning employees to different processes to maximise their value when bots replace employees. *"When you have sold out your processes saying these are the benefits [of RPA], they completely eliminate those resources from that process itself, not from the organisation, I would say from that process, and they try to deploy those resources onto the other processes. So that's where the biggest problem lies because now you don't have a dedicated resource working on that particular process"* (I1). As a result of employee redeployment, there is a possibility that no one in the team possesses end-to-end process knowledge due to reduced hands-on involvement. *"...later down the line when there are so many iterations of employees going here and there, nobody on the business side is 100% aware of the process, how it works, or how it was working before so that...knowledge gradually reduces at either one point it might disappear"* (I7). Consequently, PKL was highlighted as a major risk that resulted from employee redeployment due to RPA.

RPA governance was highlighted by all participants as an essential aspect of ensuring comprehensive knowledge management across RPA initiatives. A strong emphasis was placed on documentation as a mechanism to tackle the issue of PKL materialising within the context of RPA. *"With strong documentation of each process...anyone who is given the documentation can easily understand how the process has been defined... We are mitigating it [PKL] with those factors, but how much of our documentation we have hands-on experience with, you can't beat that"* (I7). In organisations with an established CoE for RPA governance, documentation standards and the maintenance of process knowledge are ensured by the CoE. *"When we make a change actually, we document that...It's the standard set by the CoE"* (I3). Accordingly, RPA governance was discussed as a positive factor associated with PKL that ensures the creation, maintenance, and accessibility of knowledge for all parties involved in RPA.

Task division was referred to as the allocation of tasks between employees and bots. RPA-related PKL was observed both in end-to-end automation (e.g., I1), and in task-level automation, where processes were hybridised among employees and bots (e.g., I2). *"PKL is happening because you're not working on the process anymore. If you work on something every day. You keep remembering it"* (I1). *"When we go back and ask business, they say, nobody in our team has done this task for the last two years. So, the knowledge within the core team, which should be the expertise of their process, is missing"* (I2). Furthermore, PKL depends on the number of tasks assigned to bots. *"I personally think it actually challenges knowledge issues. It depends on the degree of automation of tasks as well"* (EI4-RB). Participants revealed that a fragmented process due to task division could scatter process knowledge across employees and bots, reducing the knowledge retained by employees. *"As we proceed with RPA, it [process knowledge] will get further scattered. So that knowledge, that process knowledge, as it is getting scattered and scattered as we go along. There could be a negative impact on knowledge down the line"* (I7). As a result, task division was highlighted as a factor that negatively impacts RPA-related PKL.

Maintaining **process expertise** was emphasised to bring positive results for knowledge dissemination across business process re-engineering initiatives using RPA. *"Without the business process experts and their expertise, we are actually blind because, without their support and their cooperation and their assistance, we will be unable to re-engineer business processes that can be automated through RPA"* (I7). Some participants argued that process expertise could be documented and preserved (e.g., I6), but others argued that it could not be fully captured and documented because it implicitly resides in employees (e.g., I4). *"I don't think the process knowledge will be lost even if experts leave as we have already documented it"* (I6). *"We haven't captured all of the information about the process, in which case you obviously need a human in that which tells you that the loss is very minimal in terms of knowledge"* (I4). Accordingly, participants highlighted that process expertise is a factor that positively impacts PKL.

Combining systematic methods of BPM with RPA (**RPA-BPM integration**) was revealed to have a positive impact on RPA-based knowledge management. Some participants highlighted that, for RPA-related knowledge management, methods in BPM can be used for stakeholder engagement when monitoring and maintaining processes. *"I think COVID has taught us a lot...We should have some process in place where businesses or stakeholders should be engaged throughout the journey, even though you have handed over the bot to them... We tell our BAs [business analysts], you go back to the business, make them understand what they need to do."* (I1). Other participants highlighted that BPM and RPA use similar methods while showing the potential to integrate those. *"That actually is a business rule that needs to be incorporated in BPMS because they're doing much more than just one step...I would assume bots are also geared toward doing something similar. Methods are pretty much the same"* (I4). Likewise, RPA-BPM integration was suggested as a means of overcoming PKL.

Continuous process redesign was identified as an additional theme that emerged through primary data. This theme refers to ongoing and iterative improvements that are made to automated processes, adapting them to changing needs and desired outcomes. *"This new security regulation came in. Previously, you only needed to do a single ID check with just a single driving license number. Now, you will be required to enter the card number. So that's the process change, so the board needs to perform additional extra steps to input new data to satisfy that security or regulatory impact"* (IN4). Participants highlighted that when an end-to-end RPA-integrated process undergoes frequent changes, it might challenge employees to stay updated and gain a comprehensive understanding of the updated process. This happens because employees are not as involved in executing a process while bots that are in place undergo rigorous changes to conform with the updated workflows. Consequently, the risk of PKL increases. *"In RPA, how it [PKL] is getting impacted because now a bot does most of the tasks, and you are not aware of these process changes most of the time. In RPA, these changes are rigorous and align with modifying a bot frequently"* (IN4).

5 Discussions

In this section, we present a conceptual model with factors impacting RPA-related PKL derived from the primary interview data and secondary data synthesis from the literature. This is followed by a brief discussion about limitations.

5.1 A Conceptual Model for RPA-Related PKL

Figure 2 depicts ten factors and their positive/negative impact on RPA-related PKL. All factors are classified under three themes, namely, human factors (HF), organisational factors (OF), and process factors (PF).

We identified four factors, namely, employee turnover (HF1), knowledge hiding (HF2), automation complacency (HF3), and continuous process redesign (PF5), that contribute to or negatively impact the phenomenon of RPA-related PKL. **Employee turnover** is highlighted as a common occurrence that also has a negative impact on the employees' understanding of processes in the context of RPA. In line with the literature [38,39], primary data revealed that **knowledge hiding** impacts formal and informal knowledge networks (e.g., social groups) within the organisation. Interview findings confirmed **automation complacency** or high reliance on RPA contributes to an employee's reduced understanding of the overall process, which leads to an inability to accurately execute the process in the absence of a bot. **Continuous process redesign** was discussed as a factor that increases the risk of PKL as frequent changes are introduced to processes integrated with RPA.

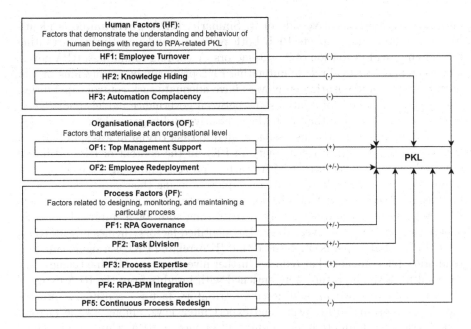

Fig. 2. A conceptual model with factors impacting RPA-related PKL

There are three factors that positively impact RPA-related PKL, namely, top management support (OF1), process expertise (PF3), and RPA-BPM integration (PF4). According to the empirical study of [31], the senior management acknowledges the employees' concerns and invests in training and internal roadshows to ensure efficient human-machine collaboration. Likewise, participants emphasised that **top management support** facilitates continuous knowledge management and upskilling of employees who are replaced by bots. Furthermore, both primary and secondary data confirmed that **process expertise** facilitates process knowledge visibility and exchange within organisations. **RPA-BPM integration** was discussed as an approach to optimise BPM's capabilities and insights with RPA to positively influence knowledge management in RPA projects [23]. As per the primary data, integrating systematic BPM methods with RPA can facilitate knowledge management when monitoring and maintaining RPA-integrated processes. Accordingly, the existing relationship between RPA-BPM integration and PKL is substantiated based on the primary and secondary data.

There are three factors, namely, employee redeployment (OF2), RPA governance (PF1), and task division (PF2), that appear to vary across different organisations, demonstrating both positive and negative impacts on RPA-related PKL. **Employee redeployment** contributes to a reduction of end-to-end process knowledge within the team, followed by reduced human involvement in the automated process. Employee redeployment, however, has often been linked to the retention of critical knowledge within organisations. While RPA literature agreed with the positive outcomes of employee redeployment, empirical data stood on the negative side of it. Similarly, **RPA governance** is identified as a factor that impacts PKL both positively and negatively, warranting further empirical validation of its sub-factors. According to findings, RPA governance can play a critical role in mitigating PKL primarily by maintaining proper documentation and ensuring employees have adequate task control capabilities. However, the evidence of relationships between governance models and PKL is limited to the literature that shows both positive and negative links to the phenomenon. In literature, the discussion related to **task division** was split between having positive and negative impacts on PKL in the context of RPA. Primary data indicated that the use of RPA scatters process knowledge due to process fragmentation, eventually resulting in employees losing their end-to-end process knowledge. Accordingly, task division was identified as a factor that impacts PKL both positively and negatively, necessitating additional empirical research.

Overall, the analysis revealed how the identified factors/themes impact RPA-related PKL. Participants emphasised that RPA-related PKL is a significant contemporary phenomenon that requires further investigation. All nine themes that emerged from the literature were enriched with insights from the RPA experts. Primary data revealed an additional contributory factor - continuous process redesign. According to literature and expert interviews, among all the factors, task division was found to have a significant impact on RPA-related PKL. Thus, focusing on task division when developing strategies may help organisations to significantly mitigate RPA-related PKL. Furthermore, primary data showed that RPA-related PKL is present in both task-level and end-to-end automation.

5.2   Study Limitations

There are several limitations of this study. First, the literature review was based on academic publications and white papers. However, the white papers lack academic rigor limiting the quality of insights. Second, the search criteria might be incomplete. For example, keywords used to search literature might have not covered all related terms as some papers may have omitted or not referred to the search terms used in this study in their keywords or abstract. Thus, these papers might not have been included in the literature review. Third, the scope of the literature review was limited to low-code automation to extract the most relevant literature related to the phenomenon under study. However, literature related to other forms of BPA has not been considered in this study. Fourth, the current conceptual model was developed based on the findings in RPA literature, along with expert insights. As RPA experts were originally from different industries with varying years of experience in the context of RPA, evidence might reflect difficulties in capturing a holistic view of the phenomenon. One such example is an RPA expert from a technical background who may approach the phenomenon under study strongly from purely a technical perspective. As a result, their responses might not be as well-rounded, reflecting all perspectives of the phenomenon. Fifth, despite adhering to a critical review process, data analysis may have been subjected to researchers' bias due to their varying backgrounds.

6   Conclusion and Future Work

This study investigated how RPA impacts PKL. RPA-related PKL is becoming an area that demands scholarly attention, as it adversely impacts an organisation's continuous improvement efforts, competitive advantages, and productivity. There are very few empirical studies specifically focusing on PKL in the context of RPA. Thus, to date, the area remains largely unexplored. The existing studies are also limited to identifying PKL as a process-related or knowledge-management challenge in RPA. We addressed this research gap by conducting a comprehensive literature review followed by seven RPA expert interviews. The study's most important finding is that PKL exists in the context of RPA. We identified ten factors that can impact the phenomenon either positively or negatively, out of which one factor emerged as an additional theme through primary data. Accordingly, this study primarily distinguishes itself from other related studies in that it identifies empirically validated factors that specifically impact RPA-related PKL. This research also signals to future researchers the existence of negative consequences of the use of RPA technology which require further investigation. Furthermore, PKL can be mitigated if organisations invest time and resources in improving the factors that positively impact the phenomenon, such as process expertise and RPA-BPM integration. Likewise, organisations can use these findings to develop strategies in line with the inherent outcome of each factor on RPA-related PKL to potentially mitigate the negative effects.

The existing conceptual model is empirically validated with rich insights from RPA experts. In future work, an exploratory case study will be conducted to further refine the model in terms of identifying the interrelationships among these factors. Exploring the relevant phenomenon in the context of an organisation will be beneficial to further examine the contextual nuances of factors within a homogeneous setting. Furthermore, we will conduct multiple case studies to enrich these insights and use cross-case analysis to construct a theoretical framework for explaining RPA-related PKL.

References

1. Adler, P.S.: Automation and skill: new directions. Int. J. Technol. Manage. **2**(5–6), 761–772 (1987)
2. Alcover, C.M., Guglielmi, D., Depolo, M., Mazzetti, G.: "Aging-and-tech job vulnerability": a proposed framework on the dual impact of aging and AI, robotics, and automation among older workers. Organ. Psychol. Rev. **11**(2), 175–201 (2021)
3. Amaravadi, C.S., Lee, I.: The dimensions of process knowledge. Knowl. Process. Manag. **12**(1), 65–76 (2005)
4. Asatiani, A., Penttinen, E.: Turning robotic process automation into commercial success – case OpusCapita. JIT Teach. Cases **6**(2), 67–74 (2016)
5. Asatiani, A., Penttinen, E., Rinta-Kahila, T., Salovaara, A.: Implementation of automation as distributed cognition in knowledge work organizations: six recommendations for managers. In: ICIS, vol. 7 (2019)

6. Asatiani, A., Penttinen, E., Ruissalo, J., Salovaara, A.: Knowledge workers' reactions to a planned introduction of robotic process automation—empirical evidence from an accounting firm. In: Hirschheim, R., Heinzl, A., Dibbern, J. (eds.) Information Systems Outsourcing. PI, pp. 413–452. Springer, Cham (2020). https://doi.org/10.1007/978-3-030-45819-5_17

7. Bandara, W., Furtmueller, E., Gorbacheva, E., Miskon, S., Beekhuyzen, J.: Achieving rigor in literature reviews: insights from qualitative data analysis and tool-support. Commun. Assoc. Inf. Syst. **37**(8), 154–204 (2015)

8. Cewe, C., Koch, D., Mertens, R.: Minimal effort requirements engineering for robotic process automation with test driven development and screen recording. In: Teniente, E., Weidlich, M. (eds.) BPM 2017. LNBIP, vol. 308, pp. 642–648. Springer, Cham (2018). https://doi.org/10.1007/978-3-319-74030-0_51

9. Costin, B.V., Anca, T., Dorian, C.: The main benefits-risks of adopting robotic process automation in big four companies from Romania - a case study. In: ICSTCC, pp. 340–345 (2021)

10. Daghfous, A., Belkhodja, O., Angell, L.C.: Understanding and managing knowledge loss. J. Knowl. Manag. **17**(5), 639–660 (2013)

11. Davenport, T.H., Prusak, L.: Working knowledge: how organizations manage what they know. Harvard Business Press (1998)

12. Dumas, M., Rosa, M.L., Mendling, J., Reijers, H.A.: Fundamentals of Business Process Management, 2nd Edn. Springer, Heidelberg (2018). https://doi.org/10.1007/978-3-662-56509-4

13. Eikebrokk, T.R., Olsen, D.H.: Robotic process automation and consequences for knowledge workers; a mixed-method study. In: Hattingh, M., Matthee, M., Smuts, H., Pappas, I., Dwivedi, Y.K., Mäntymäki, M. (eds.) I3E 2020. LNCS, vol. 12066, pp. 114–125. Springer, Cham (2020). https://doi.org/10.1007/978-3-030-44999-5_10

14. Eulerich, M., Waddoups, N., Wagener, M., Wood, D.A.: The dark side of robotic process automation (rpa): Understanding risks and challenges with RPA. Account. Horiz. **38**(1), 1–10 (2023)

15. Figueiredo, A.S., Pinto, L.H.: Robotizing shared service centres: key challenges and outcomes. J. Serv. Theory Pract. **31**(1), 157–178 (2021)

16. Forrester consulting: the forrester wave™: robotic process automation, Q2 2018. Tech. rep., Forrester Consulting (2018). https://www.uipath.com/hubfs/The_Forrester_Wave_RPA_2018_UiPath_RPA_Leader.pdf

17. Gernreich, C.C., Bartelheimer, C., Wolf, V., Prinz, C.: The impact of process automation on manufacturers' long-term knowledge. In: ICIS (2018)

18. Hallikainen, P., Bekkhus, R., Pan, S.: How OpusCapita used internal RPA capabilities to offer services to clients. MIS Quart. Execut. **17**(1), 41–52 (2018)

19. Herm, L., Janiesch, C., Helm, A., Imgrund, F., Hofmann, A., Winkelmann, A.: A framework for implementing robotic process automation projects. Inf. Syst. e-Bus. Manag. **21**(1), 1–35 (2023)

20. Hofmann, P., Samp, C., Urbach, N.: Robotic process automation. Electr. Markets **30**(1), 99–106 (2020). https://doi.org/10.1007/s12525-019-00365-8

21. de Holan, P.M., Phillips, N., Lawrence, T.B.: Managing organizational forgetting. MIT Sloan Manag. Rev. **45**(2), 45–51 (2004)

22. Kedziora, D., Penttinen, E.: Governance models for robotic process automation: the case of Nordea bank. JIT Teach. Cases **11**(1), 20–29 (2020)

23. König, M., Bein, L., Nikaj, A., Weske, M.: Integrating robotic process automation into business process management. In: Asatiani, A., et al. (eds.) BPM 2020. LNBIP,

vol. 393, pp. 132–146. Springer, Cham (2020). https://doi.org/10.1007/978-3-030-58779-6_9

24. Lacity, M.C., Willcocks, L.P.: Robotic process automation at telefonica O2. MIS Quart. Execut. **15**(1), 21–35 (2016)

25. Looy, A.V.: A quantitative and qualitative study of the link between business process management and digital innovation. Inf. Manag. **58**(2), 103413 (2021)

26. Marciniak, P., Stanisławski, R.: Internal determinants in the field of RPA technology implementation on the example of selected companies in the context of industry 4.0 assumptions. Information **12**(6), 222 (2021)

27. Meironke, A., Kuehnel, S.: How to measure RPA's benefits? A review on metrics, indicators, and evaluation methods of RPA benefit assessment. In: Wirtschaftsinformatik 2022 Proceedings, vol. 5 (2022)

28. Modliński, A., Kedziora, D., Jiménez Ramírez, A., del Río-Ortega, A.: Rolling back to manual work: an exploratory research on robotic process re-manualization. In: Marrella, A. et al. (eds.) Business Process Management: Blockchain, Robotic Process Automation, and Central and Eastern Europe Forum. BPM 2022. LNCS, vol. 459, pp. 154–169. Springer, Cham (2022). https://doi.org/10.1007/978-3-031-16168-1_10

29. Noppen, P., Beerepoot, I., van de Weerd, I., Jonker, M., Reijers, H.A.: How to keep RPA maintainable? In: Fahland, D., Ghidini, C., Becker, J., Dumas, M. (eds.) BPM 2020. LNCS, vol. 12168, pp. 453–470. Springer, Cham (2020). https://doi.org/10.1007/978-3-030-58666-9_26

30. Oshri, I., Plugge, A.: What do you see in your Bot? Lessons from KAS bank. In: Oshri, I., Kotlarsky, J., Willcocks, L.P. (eds.) Global Sourcing 2019. LNBIP, vol. 410, pp. 145–161. Springer, Cham (2020). https://doi.org/10.1007/978-3-030-66834-1_9

31. Oshri, I., Plugge, A.: Introducing RPA and automation in the financial sector: lessons from KAS bank. J. Inf. Technol. Teach. Cases. **12**(1), 88–95 (2022)

32. Plattfaut, R., Borghoff, V.: Robotic process automation: a literature-based research agenda. J. Inf. Syst. **36**(2), 173–191 (2022)

33. Plattfaut, R., Borghoff, V., Godefroid, M., Koch, J., Trampler, M., Coners, A.: The critical success factors for robotic process automation. Comput. Ind. **138**, 103646 (2022)

34. Rehr, D., Munteanu, D.: The promise of robotic process automation for the public sector. Tech. rep., George Mason University Center for Business Civic Engagement (2021). https://cbce.gmu.edu/wp-content/uploads/2021/06/The-Promise-of-RPA-For-The-Public-Sector.pdf

35. Ruiz, R.C., Ramírez, A.J., Cuaresma, M.J.E., Enríquez, J.G.: Hybridizing humans and robots: an RPA horizon envisaged from the trenches. Comput. Ind. **138**, 103615 (2022)

36. Saldana, J.: The coding manual for qualitative researchers. SAGE (2021)

37. Seethamraju, R., Marjanovic, O.: Role of process knowledge in business process improvement methodology: a case study. Bus. Process. Manag. J. **15**(6), 920–936 (2009)

38. Serenko, A.: The great resignation: the great knowledge exodus or the onset of the great knowledge revolution? J. Knowl. Manag. **27**(4), 1042–1055 (2023)

39. Serenko, A.: The human capital management perspective on quiet quitting: recommendations for employees, managers, and national policymakers. J. Knowl. Manag. (2023). https://doi.org/10.1108/JKM-10-2022-0792

40. Siderska, J.: The adoption of robotic process automation technology to ensure business processes during the COVID-19 pandemic. Sustainability **13**(14), 8020 (2021)
41. Stravinskienė, I., Serafinas, D.: Process management and robotic process automation: the insights from systematic literature review. Organ. Vadyba Sist. Tyrimai **85**(1), 87–106 (2021)
42. Syed, R., et al.: Robotic process automation: Contemporary themes and challenges. Comput. Ind. **115**, 103162 (2020)
43. Syed, R., Wynn, M.T.: How to trust a Bot: an RPA user perspective. In: Asatiani, A., et al. (eds.) BPM 2020. LNBIP, vol. 393, pp. 147–160. Springer, Cham (2020). https://doi.org/10.1007/978-3-030-58779-6_10
44. Thompson, J.: A guide to abductive thematic analysis. Qual. Rep. **27**(5), 1410–1421 (2022)
45. Zelenka, M., Vokoun, M.: Information and communication technology capabilities and business performance: the case of differences in the Czech financial sector and lessons from robotic process automation between 2015 and 2020. Rev. Innov. Compet. **7**(1), 99–116 (2021)
46. Zhang, C.A., Thomas, C., Vasarhelyi, M.A.: Attended process automation in audit: a framework and a demonstration. J. Inf. Syst. **36**(2), 101–124 (2022)

Business Models of Robotic Process Automation

Eva Katarina Helbig[1]([✉]) and Simone Braun[2] [iD]

[1] Burda Digital Systems GmbH, Hubert-Burda-Platz 1, 77652 Offenburg, Germany
eva.katarina.helbig@burda.com
[2] Offenburg University of Applied Science, Badstraße 24, 77652 Offenburg, Germany
simone.braun@hs-offenburg.de

Abstract. Robotic Process Automation (RPA) is a technology for automating business processes and connecting systems by means of software robots in organizations that is gaining traction and growing out of its infancy. Thus, it is no longer just a question of what is technologically feasible, but rather how this technology can be used most profitably. However, business models for RPA remain underinvestigated in literature. Existing work is highly heterogenous, lacking structure and applicability in practice. To close this gap, we present an approach to sustainably establish RPA as a driver of digitization and automation within a company based on an iterative, holistic view of business models with the Business Model Canvas as analysis tool.

Keywords: Robotic Process Automation · RPA · Software Robots · Process Automation · Business Model · Business Model Canvas

1 Introduction

RPA offers a seemingly lightweight way to automate (parts of) processes in an organization and link systems using rules-based software robots that mimic the humans' interaction in order to relieve employees from tedious, repetitive tasks [1]. The software robots can execute the previously defined activities 24/7 if the data and activities to be used are available in a digitized, structured, and rule-based manner [2]. It leverages graphical user interfaces (GUIs) to seamlessly bridge system gaps in the absence of Application Programming Interfaces (APIs). This eliminates the need for modifying existing applications and allows for the integration of missing interfaces, effectively bridging system breaks [2]. RPA technology is industry- and application-neutral and often implemented around back-office applications, in human resources, finance and accounting, or where large developments would be too costly [3]. RPA projects are considered being implemented quickly [4].

According to a survey of more than 1.000 companies worldwide conducted in 2021, RPA was the application most frequently used for processes [5]. As RPA becomes widespread, it is no longer just about the selection of the right RPA software but rather how the technology can be used in a way that maximizes the return on investment. Studies looking into the implementation of RPA show that many RPA projects fail because of

J. Köpke et al. (Eds.): BPM 2023, LNBIP 491, pp. 89–105, 2023.
https://doi.org/10.1007/978-3-031-43433-4_6

non-technical issues rather than technical challenges [6]. Thus, it's necessary to consider a combination of strategic, organizational, and cultural challenges to achieve long-term success [7]. Part of the strategic challenges is the transformation of the business model through RPA implementation. However, there is a lack in RPA investigation within the context of business models [8, 9]. Publications on business model development using RPA exhibit significant heterogeneity, resulting in a lack of structure and limited practical applicability. This study aims to gain a holistic view of RPA business models in practice by exploring the following research question: "*How are RPA business models structured in practice?*" We propose business models of RPA specifically relevant to practitioners who want to establish RPA in the corporate environment in the long term and in the role of the user, rather for the own use of RPA or selling the service of developing use cases with RPA.

Starting point of this research is the case of a German group-internal IT service provider concerned with the development of a new line of business: RPA. The goal is to place this technology within the company but facing the gap between theory and practice regarding establishing a business model of RPA. To meet the challenges, interviews with experts are conducted to gain a holistic understanding of the interaction between a wide range of strategic, organizational, and technical factors. The Business Model Canvas (BMC) is used as a guide [10].

This paper is structured as follows: Sect. 2 provides a brief overview of the theory of business models and the BMC and looks at best practices of business models in the field of RPA. Section 3 presents the research methodology and Sect. 4 the research results and business models of RPA. Followed by expert interviews to gain practical knowledge, the results are faced in Sect. 6, subdivided into the nine components of the BMC. Section 7 introduces a case study to validate the proposed RPA business model canvas. Section 8 discusses the results, the contribution to theory and practice, and provides an outlook.

2 Foundations of Business Models and RPA

2.1 Business Models and Business Model Canvas

Osterwalder and Pigneur [10] define the business model as "the rationale of how an organization creates, delivers, and captures value". The authors describe it through the utilization of the BMC, which serves as a conceptual framework consisting of nine building blocks that are carefully examined in relation to one another. It can be understood as a link between strategy and business processes [11]. The widely adopted BMC offers the opportunity to visually break down a complex business model into its key components, enabling a structured representation that serves as a foundation for analysis, modifications, optimization, and holistic depiction of the business model [12]. The nine key components are: (1) the value proposition (VP), (2) customer segments (CS), (3) customer channels (CC), (4) customer relationship (CR), (5) key resources (KR), (6) key activities (KA), (7) key partners (KP) as well as (8) revenue streams (RS) and (9) cost structure (CST). These form together the value creation, value delivery, and value capture [13].

The value proposition stays at the center of the BMC. Businesses are responsible for addressing the interests, requirements, and desires of their customers by providing a distinct value proposition and resolving their problems. It's essential to identify and highlight the Unique Selling Proposition (USP) that differentiates the organization from its competitors. Building blocks (2)–(4) collectively form the value delivery aspect that takes the customer perspective into account and showcases how the created value can be effectively delivered to customers. It deals with customers who need to be positioned at the core of a business model. The target group is defined, channels are determined to facilitate the purchase and assessment of the value proposition, and the relationship with each customer is cultivated. The primary focus is on the relationship that customers expect and desire, rather than the one the organization prefers and seeks to define. Building blocks (5)–(7) contribute to the value creation that sheds light on the value generation system of an organization. Key resources are indispensable for value generation and can encompass tangible and physical goods, as well as financial, human, and intellectual resources. Key activities involve tasks related to value creation and problem-solving. The network of suppliers and supportive individuals is defined as part of the key partners. Finally, building blocks (8) and (9) address the value capture, which revolves around the revenue model. Revenue is generated once the target audience is willing to pay for the offered value. Costs encompass monetary expenses necessary for resource provision, establishing a supportive network, and conducting activities [11, 14].

2.2 Business Models in the Field of Robotic Process Automation

The development of a working, profitable and sustainable business model depends on a multitude of factors. For example, Axmann et al. [15] analyzed and categorized cost drivers and proposed a framework for estimating the costs of RPA projects. Besides of costs, multitude factors include, e.g., company size, organizational structure, and strategy. Asatiani et al. [16] examined questions like how RPA should be deployed and how to build an optimal Operating Model for the company. This goes beyond mere cost and benefit comparison, which is already a challenge to be measured for RPA according to [17]. Factors such as digital mindset within the company and the motives for moving towards RPA must be considered as well. Thus, there is no universal solution for building a consistent business model for RPA [18]. Plogmaker et al. [19] embedded RPA in a business model to include technical, economic, as well as organizational and social aspects, such as customer benefit, purchasing or strategy with a focus on value creation.

3 Research Methodology

Our research methodology follows the Design Science Research approach [20] adapted by [21] consisting of four phases: (1) awareness of problem, (2) data collection and suggestion, (3) development, and (4) evaluation and conclusion.

Awareness of problem: Starting point is the case of a large German company's internal IT service provider developing a new line of business, RPA. The aim is to implement the technology but faced challenges in creating a sustainable business model. After an initial literature search and in accordance with [8, 9], existing work and practice-oriented publications lack explanations of integrating RPA into the business and achieving goals. Transferability and access to general insights for drawing conclusions about individual applicability are limited [22]. There is a disparity between theory and practice when it comes to building an RPA business model.

Data collection and suggestion: To identify the empirical values and best practices associated with RPA and business models of individual companies, an extended literature review is conducted following [23]. For this purpose, the literature review starts by defining superordinate keywords to narrow down the search. First, a keyword query is performed on Google Scholar, searching mainly for the keyword Robotic Process Automation, RPA, and Business Model. We include English and German literature (with corresponding keyword translations). Further refinement of the search is done by deriving more specific keywords, such as process automation, process optimization, standardization, Business Model Canvas, and software robots, after an initial review of scientific literature. Various online library catalogs and databases are assessed, including those of the university, Springer Link, and ScienceDirect. To answer the research question in Sect. 1, specific criteria are set for the literature, like the use of technical literature, scientific publications and studies, references to RPA and business models or cross-cutting topics, and high-quality scientific publications. From the analysis of the literature, we extract seven distinct categories of crucial factors for creating a business model of RPA. In a second step, we conduct semi-structured expert interviews [24] using the BMC to qualitatively analyze and interpret the gathered data. The aim of this application-oriented research is to generate practical knowledge for addressing real-world problems by looking at the entirety of a holistic business model, its building blocks and how they relate to each other [25].

Development: The findings of the expert interviews are compiled, taking into consideration the theoretical foundations and the current state of research, thereby creating a synthesis between theory and practice. They are organized into Sections on value proposition, value delivery, value creation, and value capture. These empirical results, extracted from the experiences, are consolidated within a BMC.

Evaluation and conclusion: To verify the theory in practice and validate statements while gaining new practical, realistic insights, a single-case study [26] is conducted using Burda Digital Systems GmbH as an example. The case study consists of expert interviews and exploratory process development, resulting in a PDCA-list (Plan-Do-Check-Act) within the company [27].

4 Literature Review

Real-world examples came from the energy, finance, and pharmaceutical sectors [22]. It is crucial to consider these aspects collectively, as they are interrelated. We organized them into seven distinct categories, presented in the following (see Table 1).

(I) Organizational: To anchor RPA within an organization, it's necessary to establish a multidisciplinary Center of Excellence (CoE) with process analysts, developers, architects, and managers as key roles [22].

(II) Operating Model: A central part of the RPA deployment is the operating model, including the individual units and their responsibilities. We can differentiate three different forms: the centralized model, the decentralized model, and the hybrid model. The centralized model includes a CoE, usually situated within the IT unit. Lines of communication are short, scaling is easy, all knowledge is bundled in one central location

and it reduces the risk of multiple RPA solutions running separately and in parallel [22]. The decentralized model includes many small CoE, located in the various business units. They are structured differently depending on the organization. The hybrid model is characterized by a CoE with additional process analysts in the individual units [18].

(III) Development Approach: A CoE may pursue different development approaches, such as the make, buy or offshore approach. With the buy approach, external developers implement processes independently or in cooperation with internal developers. With the make approach, the organization itself provides the developers as a resource. With the offshore approach, the development of the processes is partially or completely handed over to developers at offshore locations, who in turn pass the productive implementation of the processes to the onshore locations. This approach seems inexpensive, but has the disadvantage of the lack of process knowledge and the lack of corporate mentality on the part of the offshore developers [16].

(IV) Change Management: Change Management focuses on successfully establishing a new RPA technology in the organization involving employees and managers in this process [6]. The focus must be on transparent communication towards the staff and conveying strategy and vision in the interest of the individual well-being of the employees. The introduction of RPA technology can be established through a top-down or bottom-up approach. With the top-down approach, RPA is controlled from the executive level without prior consultation or communication with employees, often leading to negative and destructive attitudes towards RPA. With the bottom-up approach, technology is introduced or accompanied by the employees themselves, resulting in better understanding and enthusiasm. This successful integration can be supported if employees identify new processes for implementation themselves and thus endorsing the new technology [22, 28].

(V) Service Model: When choosing the service model, there is the option of using RPA as Software-as-a-Service (SaaS) with monthly license fees for using the software. In addition, there is the Automation-as-a-Service (AaaS) in the RPA environment. With the AaaS the service provider is responsible for the control and administration of the software robots as well as the development, operation, and monitoring of the processes. The software robots are running in the own data center. This option basically serves as an alternative to setting up an own RPA business model [16].

(VI) Security Concept and Monitoring: Since software robots work with data, certain compliance and data protection guidelines must be adhered to. It is recommended to set up an internal set of rules with guidelines, risks, and mitigation actions. In addition, monitoring plays a crucial role in process monitoring and measuring performance [18, 29].

(VII) Infrastructure and Pipeline- and Case-Management: To ensure testability, traceability, scalability, and robustness, the use of development, stage, and production environments is recommended. A critical aspect is the identification and prioritization of processes that yield significant benefits [18].

Table 1. Factors for creating a business model of RPA.

Category	Description	Ref
(I) Organizational	• Establish a multidisciplinary CoE • Process Analyst: identification, analysis, documentation, organization, structuring, process capture • Developer: design, development, testing, documentation, maintenance, process evolution • Architect: rollout, support, enhancement, and scaling of the RPA software • Manager: strategy, vision, point of contact, monitoring	[16, 18, 30]
(II) Operating Model	• Centralized model: centralized CoE • Decentralized model: various CoE in the business units • Hybrid model: centralized CoE + process analysts in units	[16, 18]
(III) Development Approach	• Buy approach: external developers • Make approach: internal developers • Offshore approach: developers at offshore locations (lack of process knowledge)	[16]

(IV) Change Management	• Involvement of employees and managers
	  – Top-down approach: introduction of the technology starts from the leadership level; Bottom-up approach: introduction of the technology starts from the employee level
	• Integration of the technology within the organization
	  – Identification of processes by employees
	  – Communication of the company's motivations
	  – Establishment of a shared vision
	• Development of a strategy
		[29, 30]

(continued)

Table 1. (*continued*)

Category	Description	Ref
(V) Service Model	• Software-as-a-Service: license fee for cloud-based software usage • Automation-as-a-Service: cloud-based control and administration (service provider takes responsibility for the development, operation, and monitoring), software robots are hosted in the customer's data center (offering an alternative to building an in-house RPA model)	[16]
(VI) Security & Monitoring	• Compliance and data privacy • Documentation of guidelines, risks, and measures • Process ownership retained within the department • Quantification of RPA performance (KPIs)	[18, 29, 31]
(VII) Infrastructure & Pipeline- and Case-Management	• Use of development, stage, and production environment • Identification and prioritization of new processes • Managing customer demands • Implementing a back log for improvements (6-month cycle)	[17, 18]

5   Expert Interviews

A semi-structured approach is adopted for the expert interviews with an interview guide to ensure data comparability [24]. The BMC is used as analysis tool to support structuring and focusing on the aspects of value proposition, creation, delivery, and capture. A set of 25 potential questions is identified, shown in Table 2, for 60-min interviews via Microsoft Teams, with audio recordings for data collection. Interview duration varied between 20 to 75 min due to time constraints and unanswered questions. Interviews are anonymized to foster trust and openness. Data processing entails organizing and structuring information for preparation and evaluation purposes. The transcription includes the rule-based assignment of the answers to the questions and the building blocks of the BMC. The transcription by meaning is a sufficient means for eliciting descriptive and subjective experiences of the interviewees.

Table 2. Interview guide.

(I) Personal		1. What is the scope of your position?
(II) Value Proposition		2. What problems do your customers face and how do you solve them? 3. How long have you been offering this service?
(III) Value Delivery	*(a) KR*	4. Which software do you use and why? 5. Is there any additional RPA software you are using? 6. Please describe your setup 7. How many people are dealing with the topic of RPA within the company? 8. How do you train yourself and your employees?
	(b) KA	9. What activities are involved?
	(c) KP	10. What collaborations are needed to provide your service?
	(d) CST	11. What are your costs?
(IV) Value creation	*(e) RS*	12. What is the composition of your revenues? 13. At what size of a process is it worth to order its automation?
	(f) CR	14. How do you manage the relationship with customers? 15. How can customers give feedback? 16. What does the first meeting with customers look like? 17. How is RPA accepted by customers, employees, and management?
	(g) CC	18. How do you market RPA? 19. Which channels do you use to interact with customers and clients?
	(h) CS	20. How do you describe your target group or which processes can be automated well?
(V) Optional Questions		21. What expectations towards RPA did you start with, how do you feel about it now? 22. What challenges do you currently face? 23. What are future goals? 24. What lessons learned have you been able to gather? 25. What do you think is key to a successful business model of RPA?

The target group comprise professionals working in the field of RPA across different industries. Since developing a business model requires a comprehensive understanding of various aspects, the aim is to gain insights into the diverse perspectives of RPA.

To achieve a well-rounded and inclusive representation, a heterogeneous mix of companies is sought, resulting in 21 interviews conducted (see Table 3). The interviewed experts serve as representatives of their organizations with valuable knowledge in the domain of RPA, offering practical insights into their subjective experiences, opinions, and attitudes. The interviewees are classified into three RPA experience levels: Beginner (<1 year), Intermediate (1–3 years), and Advanced (>3 years). Organizations included those engaged in developing *internal* processes using RPA at corporate and organizational levels as well as those focusing on providing *external* consulting and implementation of RPA solutions to clients. This highlights the value of RPA both as a product and as a service.

Table 3. Overview of the interviewees' characteristics.

Personal		Company-specific			
ID	Experiences	Industry	HQ	#Employees	Business model
A	Intermediate	Chemistry	DE	10.001 +	Internal
B	Beginner	Machine	DE	10.001 +	Internal
C	Advanced	Automobile	FR	10.001 +	Internal
D	Beginner	Finance	LU	10.001 +	Internal
E	Advanced	Biotech	US	5.001–10.000	Internal
F	Advanced	Finance	DE	5.001–10.000	internal & external
G	Intermediate	Logistics	DE	5.001–10.000	Internal
H	Intermediate	Finance	KG	501–1.000	Internal
I	Advanced	Finance	DE	501–1.000	Internal
J	Advanced	IT & Consulting	DE	501–1.000	External
K	Beginner	Real estate	DE	50–1.000	Internal
L	Beginner	IT & Consulting	DE	501–1.000	Internal
M	Intermediate	IT & Consulting	DE	201–500	External
N	Intermediate	Event	DE	201–500	Internal
O	Advanced	Chemistry	IN	51–200	Internal
P	Advanced	IT & Consulting	DE	11–50	External
Q	Intermediate	IT & Consulting	DE	11–50	Internal
R	Beginner	IT & Consulting	DE	11–50	External
S	Intermediate	IT & Consulting	DE	2–10	External
T	Beginner	IT & Consulting	DE	201–500	Internal
U	Beginner	IT & Consulting	DE	201–500	Internal

6   RPA Business Models

The expert interviews are analyzed with the integration of current theory and practice, resulting in empirical results which are organized into Sections on value proposition, delivery, creation, and capture, and finally consolidated within a BMC, as shown in Fig. 1. Selected results are summarized and discussed in the following. Please refer to Fig. 1 in cases where a specific building block is not further elaborated.

6.1   Value Proposition

Value Proposition (Building Block 1). We asked the interviewees to describe their value proposition by addressing the following two questions: What challenge does their customer base face, and how do they provide a solution? The predominant issue highlighted by the customers was a lack of time (A, C, D, F, G, S). Company D strongly affirmed this by stating "Robots do the work, we all hate" (D). Similarly, Company E emphasized the need for a straightforward resolution to the problem, noting that "most companies don't care about how it is implemented" (E). It is crucial to prioritize understanding the customers' needs and addressing their challenges (L).

6.2   Value Delivery

Customer Relationship (Building Block 4). The interviewers were asked about the nature of the relationships they maintain with their customers. Company T emphasized that relationships involve practical aspects such as competence, goal orientation, and efficiency, as well as emotional values like trust and understanding (T). To gain insights into the overall sentiment, the companies were asked about the reception of RPA by customers, employees, and management. The mood varied, but the majority expressed a highly positive sentiment (A, B, F, R). They appreciate the positive impact of RPA in eliminating tedious processes (B). Companies D, K, and L described the mood as mixed, with awareness and understanding of RPA seen as prerequisites (D, K, L). Company I initially encountered skepticism towards software robots but successfully addressed it by building a positive image and branding the robots as "Roberta" (I).

6.3   Value Creation

Key Resources (Building Block 7). During the exploration of key human resources, the interviewees were requested to provide insights into the allocation of tasks and roles related to RPA within their organizations. The distribution of roles aligns with the responsibilities outlined in Sect. 4, and the CoE plays a significant multidisciplinary role. Companies A and B follow a hybrid organizational structure, where both a CoE exists, and developers are situated within individual business units. When discussing employee training opportunities, several options were mentioned, including online communities (A, N, Q, S, T, U), knowledge exchange among colleagues (A, F), learning by doing (C, F, G, L), and assistance from external implementation partners (C, I, O). Companies C and G emphasized the importance of staying updated on their own software solutions as well as competitor technologies (C, G).

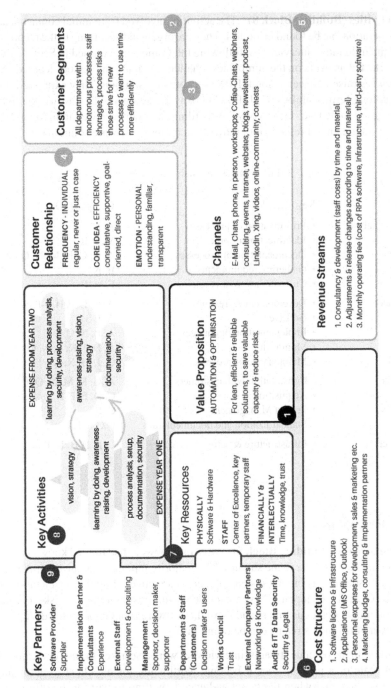

Fig. 1. RPA Business Model Canvas.

Key Activities (Building Block 8). The activities involved from initial contact to application operation can be found in Table 4. They are organized according to the responsibilities of the individual interviewees, following the framework outlined in Sect. 4. Table 5 shows selected answers on the companies' challenges, goals and lessons learned. A few interviewees raised concerns about RPA as a software solution. They expressed doubts about its long-term viability (C, J, P). Company J cautioned, "Don't trust the hype. It is a 'too good to be true' technology." According to the interviewed individual, RPA is seen as a fragile and unreliable tool rather than a robust software solution (J).

Table 4. Roles and their key activities.

Role	Key activities	Company ID
Process analysts	• Prioritization of processes • Identification of processes • Documentation of processes • Gathering requirements • Customer awareness-raising	A, C M, P, R A, C, G, K, M, N, Q - S B, N F, M, R
Developer	• Development & Testing • Go Live • Implementation of changes • Monitoring • Maintenance of processes	A, B, E, F, H, J-P, S A, I, S A, E, N, Q A, D, J, K, L, N, O G, J, M, N, Q
Manager	• Consultation with customers • Acquisition • Project Management • Standardization and documentation	A, G, K, M, N, Q, R C, D, E, F, R C, F F, G, I, N
Architects	• Infrastructure & setup	C, E, F, H, K, R

6.4　Value Capture

Revenue Streams (Building Block 5). Most interviewees had limited knowledge about the exact breakdown of revenue streams. The majority reported that the automations are not directly sold to customers; instead, the costs are covered by the organization itself (A, B, E, G, P, R). Company P even considered full cost recovery as risky because it could potentially deter customers and leave the software robot underutilized. A possible breakdown of revenue streams can be found in Fig. 1.

In line with the revenue streams, interviewees were also asked to answer the question: At what scale of a process does it become worthwhile for the client to automate it? Revenue is generated when the target audience is willing to pay for the offered value (U). Thus, investment is required in analyzing suitable processes and customers (C, D, T, U). Process identification and prioritization, as mentioned in Sect. 4, are crucial factors. Company C evaluates processes based on questions such as: How much employee capacity can be saved? Are there tasks that are not feasible without the software robot? The interviewee recommended always considering whether another application might be better suited for optimizing and solving the problem (C). Multiple companies advised starting with small processes to gain a better understanding of RPA (B, G, P, R, S). Company D provided an example: "If a process requires more than 10 days of effort but saves more than 10 days, it is personally worthwhile because employees can then perform more challenging tasks, and motivation increases" (D).

Table 5. Challenges, goals, and lessons learned.

Category	Activities
Challenges	• Profitable positioning of the software within the company (C) • Building a good strategy (C) • Sensitizing customers and employees (C, D, R)
Goals	• Integration of RPA, AI, and OCR for evaluations (D, G) • External consulting (G) • Expanding RPA competency (S)
Lessons learned	• Dare to make mistakes (A, F, M) • Integration into the company (C, L, R, S, U) • Empowering employees to identify processes (G) • Do not proceed "quick and dirty" & grow too fast (G, J) • Considering all interests, such as works council, staff, audit, and management (G, N, P, S) • Hands on transparency and awareness (L, T) • Setting the right expectations (N, P, S) • Communicating strategy and goals (I, R) • Establishing a CoE (P, R) • Plan enough time & buffers & don't grow too fast (A, C)

7   Case Study Burda Digital Systems GmbH

A case study was conducted on Burda Digital Systems GmbH (BDS) to validate the theory in practice and gain new practical insights. BDS is an internal IT service provider affiliated with the parent company Hubert Burda Media with approximately 10.500 employees worldwide. BDS acts as an automation partner for intra-group clients. The objective is to deliver prompt solutions for RPA to internal clients and establishing RPA as a sustainable and successful corporate service while building a high-performing internal organization. To that end, BDS aims at developing an effective business model of RPA and strategically expanding the use of this technology across the entire corporation.

   The results of the expert interviews and exploratory process development demonstrate similarities between theory and practice, which highlight the added value of qualitative research in practice: the acquisition of knowledge from real situations. The BMC proved to be an optimal tool for structuring and examining the business model as a whole, including the interaction of its individual building blocks. A key finding regarding the research question on the structure of an RPA business model is that a combination of various areas and aspects is crucial, such as raising awareness among employees and partners, creating an understanding of RPA, and identifying suitable processes. The developed PDCA list [27], shown in Table 6, contains potential next steps and recommendations. The content of the PDCA list is presented with a general character, detached from individual cases, and will be used, redefined, and questioned iteratively in conjunction with the BMC in the future.

Table 6. PDCA List.

Plan *Planning Phase*

- Development of a shared vision and strategy
- Involvement of employees, management, and all key partners
- Definition of the operating model (integration of RPA into the company)

Do *Implementation Phase*

- Creating transparency regarding goals, vision, and the "why"
- Raise awareness of the technology
- Standardization and optimization before automation

Check *Evaluation Phase*

- Did the process pay off?
- What would motivate existing customers to commission another process?
- How can the relationship and project stand out to the clientele?

Act *Optimization Phase*

- Capturing, sharing, and documenting lessons learned
- Creative thinking without limits: "How would we build the business model if we had unlimited resources?"
- What are the strengths and weaknesses of the current business model?

8   Discussion and Conclusions

This study contributes to a better understanding of RPA business models and addresses the gap in scientific research characterized by insufficient qualitative depth. Our findings align with existing research such as [16–18, 29–31] and further enhance it by providing a holistic view of RPA business models overcoming the heterogeneity, lack of structure, and limited practical applicability [7–9]. This was achieved through expert interviews and a case study, resulting in the creation of a uniform RPA business model canvas, which is a first in this field. This work expands the BMC by incorporating multiple perspectives and a top-down technical view.

Practitioners can learn about RPA implementation practices and gain awareness of individual components. Using the BMC as a structuring framework and creating a PDCA list for actionable recommendations, these findings help further development and transfer to their own case. Through this process, practitioners can effectively apply the current state of research on RPA business models within the analytical model framework, facilitating practical implementation.

While being widely adopted and accepted, the BMC itself has inherent limitations and alternative models should be considered and tested. It should be noted that the research findings of this study do not have a normative character that can be universally applied to others' business models. They rather offer practical insights and recommendations tailored to specific research contexts. Defining key performance indicators is advisable for measuring success after implementation.

Future work includes further validation of the results through qualitative investigations or surveys in diverse organizations and industries to identify additional criteria as well as applying and validating the created BMC. Even if the case study is successfully implemented at BDS, the BMC and the PDCA list must be continuously revised and applied iteratively. Besides concentrating on the big picture, an additional research avenue may encompass focusing on specific building blocks, considering additional models such as the Value Proposition Canvas. One goal would be the long-term stabilization of RPA in the company, including, for example, the customer segment.

In conclusion, this research examined the components of RPA business models by means of 21 expert interviews and a case study, validating the theoretical structures and processes in practical settings and providing valuable insights into the practical implementation of RPA. The gathered knowledge and recommendations are summarized in the BMC offering a comprehensive roadmap for sustainable RPA business models. The BMC highlights the importance of customer focus, technology integration, employee awareness, and process identification. It is crucial to understand that building a successful RPA business model requires more than just technological understanding. Although RPA is not as established as other technologies, it can serve as a transitional solution in the context of digital transformation. Establishing a robust RPA business model is vital to support organizations amidst the rapid changes and complexity of the digital landscape.

References

1. Eggert, M., Moulen, T.: Selektion von Geschäftsprozessen zur Anwendung von Robotic Process Automation am Beispiel einer Versicherung. HMD **57**, 1150–1162 (2020)
2. Osman, C.-C.: Robotic process automation: lessons learned from case studies. Informatica Economica **23**, 66–75 (2019)
3. Koç, H., Weisweber, W., Lüttke, M.: Robotic process automation und Digitale service-transformation bei der DRV. HMD **58**, 1230–1243 (2021)
4. Brettschneider, J.: HMD Praxis der Wirtschaftsinformatik **57**(6), 1097–1110 (2020). https://doi.org/10.1365/s40702-020-00621-y
5. Clark, J., Perrault, R.: Artificial Intelligence Index Report 2022. Stanford (2022)
6. Langmann, C.: Robotic Process Automation als Wegbereiter eines modernen Rechnungswesens. Rethinking Finan. **2**, 4–8 (2020)
7. Berghaus, S., Back, A.: Stages in digital business transformation: results of an empirical maturity study. In: MCIS 2016 Proceedings, vol. 22, pp. 1–17 (2016)

8. Sobczak, A.: Robotic process automation as a digital transformation tool for increasing organizational resilience in polish enterprises. Sustainability **14**, 1333 (2022)
9. Moreira, S., Mamede, H.S., Santos, A.: Process automation using RPA – a literature review. Proc. Comput. Sci. **219**, 244–254 (2023)
10. Osterwalder, A., Pigneur, Y.: Business Model Generation. Wiley, Hoboken (2010)
11. Kotarba, M.: Digital transformation of business models. Found. Manag. **10**, 123–142 (2018)
12. Lukas, T.: Business model canvas – Geschäftsmodellentwicklung im digitalen Zeitalter. In: Grote, S., Goyk, R. (eds.) Führungsinstrumente aus dem Silicon Valley. Springer Gabler, Heidelberg (2018). https://doi.org/10.1007/978-3-662-54885-1_9
13. Munaro, M.R., Freitas, M.d.C.D., Tavares, S.F., Bragança, L.: Circular business models: current state and framework to achieve sustainable buildings. J. Constr. Eng. Manage. **147** (2021)
14. Clark, T.: Business Model You. Wiley, Hoboken (2012)
15. Axmann, B., Harmoko, H., Herm, LV., Janiesch, C.: A framework of cost drivers for robotic process automation projects. In: González Enríquez, J., Debois, S., Fettke, P., Plebani, P., van de Weerd, I., Weber, I. (eds.) Business Process Management: Blockchain and Robotic Process Automation Forum. BPM 2021. Lecture Notes in Business Information Processing, vol. 428. Springer, Cham (2021). https://doi.org/10.1007/978-3-030-85867-4_2
16. Asatiani, A., Copeland, O., Penttinen, E.: Deciding on the robotic process automation operating model: a checklist for RPA managers. Bus. Horiz. **66**, 109–121 (2023)
17. Syed, R., et al.: Robotic process automation: contemporary themes and challenges. Comput. Ind. **115**, 1–55 (2020)
18. Fantina, R., Storozhuk, A., Goyal, K.: Introducing Robotic Process Automation to Your Organization. A Guide for Business Leaders. Apress; Safari, Boston (2022)
19. Plogmaker, H., Feldmann, C.: Geschäftsmodell-innovationen mit robotic process automation. In: Feldmann, C. (eds.) Praxishandbuch Robotic Process Automation (RPA). Springer Gabler, Wiesbaden (2022). https://doi.org/10.1007/978-3-658-38379-4_10
20. Vaishnavi, V.K., Kuechler, W.: Design Science Research Methods and Patterns. Innovating Information and Communication Technology. Taylor & Francis, Boca Raton (2008)
21. Herm, L.-V., Janiesch, C., Helm, A., Imgrund, F., Hofmann, A., Winkelmann, A.: A framework for implementing robotic process automation projects. Inf. Syst. E-Bus. Manage. **21**, 1–35 (2023)
22. Langmann, C., Turi, D.: Robotic Process Automation (RPA) - Digitalisierung und Automatisierung von Prozessen. Springer Gabler, Wiesbaden (2021)
23. Webster, J., Watson, R.T.: Analyzing the past to prepare for the future: writing a literature review. MIS Q. **26**, xiii–xxiii (2002)
24. Finkbeiner, P.: Qualitative research: semi-structured expert interview. In: Social Media for Knowledge Sharing in Automotive Repair. Springer, Cham (2017). https://doi.org/10.1007/978-3-319-48544-7_6
25. Meertens, L.O., Iacob, M.E., Nieuwenhuis, L.J.M., van Sinderen, M.J., Jonkers, H., Quartel, D.: Mapping the business model canvas to ArchiMate. In: Proceedings of SAC 2012. ACM Press, New York (2012)
26. Yin, R.K.: Case Study Research. Design and Methods. Sage, Los Angeles (2009)
27. Arredondo-Soto, K.C., Blanco-Fernandez, J., Miranda-Ackerman, M.A., Solis-Quinteros, M.M., Realyvasquez-Vargas, A., Garcia-Alcaraz, J.L.: A plan-do-check-act based process improvement intervention for quality improvement. IEEE Access **9**, 132779–132790 (2021)
28. Tang, K.N.: Change management. In: Leadership and Change Management. SpringerBriefs in Business. Springer, Singapore (2019). https://doi.org/10.1007/978-981-13-8902-3_5
29. Arnaz, R., Harahap, M.E.: Analysis of implementation of robotic process automation: a case study in PT X. In: Proceedings of the BIEC 2020. AEBMR. Atlantis Press, Paris (2020)

30. Plattfaut, R., Borghoff, V.: Robotic process automation: a literature-based research agenda. J. Inform. Syst. **36**, 173–191 (2022)
31. Hindle, J., Lacity, M., Willcocks, L., Shaji, K.: Robotic Process Automation: Benchmarking the Client Experience. Interim Executive Research Report (2017)

SiDiTeR: Similarity Discovering Techniques for Robotic Process Automation

Petr Průcha(✉) ⓘ and Peter Madzík ⓘ

Technical University of Liberec, Studentská 1402/2, Liberec, Czechia
petr.prucha@tul.cz

Abstract. Robotic Process Automation (RPA) has gained widespread adoption in corporate organizations, streamlining work processes while also introducing additional maintenance tasks. Effective governance of RPA can be achieved through the reusability of RPA components. However, refactoring RPA processes poses challenges when dealing with larger development teams, outsourcing, and staff turnover. This research aims to explore the possibility of identifying similarities in RPA processes for refactoring. To address this issue, we have developed Similarity Discovering Techniques for RPA (SiDiTeR). SiDiTeR utilizes source code or process logs from RPA automations to search for similar or identical parts within RPA processes. The techniques introduced are specifically tailored to the RPA domain. We have expanded the potential matches by introducing a dictionary feature which helps identify different activities that produce the same output, and this has led to improved results in the RPA domain. Through our analysis, we have discovered 655 matches across 156 processes, with the longest match spanning 163 occurrences in 15 processes. Process similarity within the RPA domain proves to be a viable solution for mitigating the maintenance burden associated with RPA. This underscores the significance of process similarity in the RPA domain.

Keywords: Robotic Process Automation · process similarity · RPA governance · RPA maintenance

1 Introduction

Robotic Process Automation is slowly becoming mainstream technology in various corporate organizations. Unfortunately, even though RPA makes work easier in some ways, it can generate additional work, especially during the running of RPA itself [16]. Very often this happens with companies that cross a critical threshold and fall into an RPA maintenance trap [27]. One way to prevent this, according to RPA developers, is to ensure the reusability of RPA components [7, 16, 26]. With a small number of RPA robots and a small number of RPA developers, this can be easily ensured. With a larger number of RPA robots, larger development teams, outsourcing automation to different development teams in different parts of the world and with the turnover of staff, ensuring the reusability of RPA components is very challenging. Making sure that code quality complies with company norms during development is also challenging. For this reason, software developers should refactor their code to be more efficient and serviceable.

© The Author(s), under exclusive license to Springer Nature Switzerland AG 2023
J. Köpke et al. (Eds.): BPM 2023, LNBIP 491, pp. 106–119, 2023.
https://doi.org/10.1007/978-3-031-43433-4_7

Hence, it is advisable to refactor the RPA code as well, so that the code components are reusable. As in software development, refactoring can be done backwards.

Aim of the research: *To explore the possibility of finding similarities or identical parts in an RPA process for refactoring if many automations were developed by people who no longer work in a particular company, or if the development was outsourced.*

There is an area in business process management that addresses a similar problem and then tries to find identical processes within an organization, or across manufacturing plants, or after a merger/acquisition. However, these techniques have focused on processes that are not automated. The most commonly used sources for analysis are process logs, natural language content, graph structures, Petri Nets, and BPMN notation [11, 31]. None of these methods are primarily intended for the RPA area. Therefore, input data, which for RPA may be the code of an RPA bot or possibly the log records from RPA bots, are not considered. However, using the foundation of these techniques can help answer our research question and achieve better maintainability by finding parts from RPA code to refactor into reusable components.

The need for a new similarity algorithm comes from the desire to deal with the maintenance trap. The current algorithms and solutions are not compatible with RPA processes or logs. Many current discovery techniques are discussed in the section titled Related Work. While these techniques propose interesting ideas which inspired our solution here, they would be hard to use in the RPA domain or would not be especially effective. Firstly, all currently used algorithms would need a certain amount of data preparation before their application. And then, after all of the transformations, there could arise certain problems related to the specifics of RPA technologies and the structure of process flow. For example, the process inquiry can deviate from reality. The RPA technology sometimes needs to add extra activities to the flow in order to function properly, for example exceptions which account for a loading screen. These extra activities would be problematic because in a standard graphical visualization as a BPMN or a Petri Net, these activities would not be covered. Also, the structure of the RPA code can be more problematic due to the fact that many activities are nested inside other activities. Before the analysis, it is important to flatten the process structure in order to perform an analysis. Lastly, the effectiveness of non-RPA algorithms can be lower, because in a computer environment, it is possible to perform the same action a different way and get an identical output. Our dictionary feature can recognize process activities which are different, even when the activities yield the same output. This extends the pool of similar or same activities. This increases the number of criteria for using algorithms from related work that can be used in the RPA domain after minor or major changes. These criteria will be introduced in the Related Work section.

In this article, we propose that Similarity Discovering Techniques for RPA processes shall be identified as SiDiTeRs. A SiDiTeR is a technique for searching for similar parts of RPA code which could be refactored into reusable components. A SiDiTeR is specially designed for use with UiPath RPA processes, currently the most used RPA tools [38]. The approach can be extended to other commercial RPA solutions in order to discover similarities in RPA processes. Our techniques promise to efficiently discover similar patterns in a sequence of activities to later maximize the ability to leverage the benefit of reusability of the RPA components.

The main contributions of this new algorithm for identifying process similarity in RPA processes are:

- Its ability to work on RPA designs or RPA process logs
- By design it works with the specifics of RPA technologies, like process structure and process flow
- A dictionary feature is provided to extend potential matches and cover identical outputs

In this article we first analyze the previous work related to our approach. Subsequently we describe the use of SiDiTeRs in detail as a method for RPA process similarity discovery. We follow with an evaluation of the method and a conclusion of the work.

2 Related Work

There are already other approaches for discovering process similarity. Therefore, in this section we will analyze other approaches where a discovery approach is used, what input data is needed, and also how much these approaches comply with our criteria for RPA. We assume that after tuning all of the algorithms, they could at least partly be used in the RPA domain. For example, after converting the RPA processes to another format, a certain approach could be used. For the analysis of other approaches we will classify them based on the publication on process similarity by Schoknecht et al. and Dijkman et al. [6, 29]. Most authors use more than one of these approaches to compare process similarity. Process similarity approaches are:

Behavioural similarity methods usually use execution traces of process and then analyze the change in execution states or the behaviour of the flow. That means that they check individual states and their changes.

Natural language similarity methods use natural language to try to find similarity in labels of activities. Many other approaches use both syntactic and semantic aspects of language to analyze similarity.

Graph or structural similarity methods consider graph structure or business process-aware control flow. Various techniques like the graph edit distance technique or the block structure technique are used to measure the similarity between process models based on their graph structure and control flow.

Attribute Similarity methods examine the similarity between the attributes of each activity that are required for the successful execution of that activity in the process.

The criteria for determining if related algorithms (after the necessary changes) have the ability to work effectively in the RPA domain can be summarized from the introduction of this paper. The criteria are:

1. The ability to correctly interpret RPA processes from the RPA design or an RPA log with all of the nested activities inside.
2. The ability to handle the extra activities in the RPA processes that will not be displayed in a graphical visualization of the process.
3. The ability to cover different activities with the same output.

An analysis of the criteria for a match is presented in the last column of Table 1. An analysis of related works for determining which approaches and inputs could be exploited for this study was carried out according to Fig. 1. Scopus and Web of Science databases were used to search for related works. All non-BPM records were excluded from the search results, including those from manufacturing, computer science (CPU related), databases, web services, and psychology. We also excluded works related to BPM if they were not relevant for generating similar processes or if the records were not accessible. In this eligibility screening, we also excluded records which did not provide a new method or algorithm for analyzing the process similarity or if they had not yet been validated on any processes.

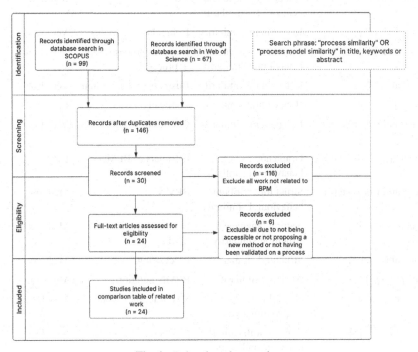

Fig. 1. Related work procedure

The result of the analysis of each approach is shown in Table 1. In Table 1 there are only the publications that passed through the filter. Our search phrases are shown in Fig. 1. Schoknecht et al. [29] conducted a similar literature review and found 123 relevant publications. However, they also used phrases and keywords which were older and, according to them, no longer used today.

Table 1. Related work comparison table

Publication	Type similarity	Format of input data	Criteria match
Ye et al. [34]	Graph similarity	Connected Graph	Low
Garcia et al. [11]	Graph similarity	(BPMN) 2.0.2	Medium
Pei et al. [25]	Behavioural similarity	Petri Net	Low
Niu et al. [24]	Behavioural similarity	Token Logs	Very Low
Liu et al. [19]	Behavioural and graph similarity	DWF-nets	Low
Sohail et al. [30]	Natural language and behavioural similarity	XML	Medium
Zeng et al. [35]	Behavioural similarity	Role relation network	Very Low
Zhou et al. [37]	Natural language and behavioural similarity	Business process graph + processLog	Very Low
Liu et al. [20]	Behavioural similarity	Business process graph	Very Low
Valero [32]	Behavioural similarity	Petri Nets	Low
Klinkmuller and Weber [18]	Behavioural similarity	Control flow log	Very Low
Cao et al. [5]	Graph and behavioural similarity	Petri nets or BPMN	Low
Amiri and Koupaee [3]	Structural, attribute behavioural similarity	BPMN	Medium
Figueroa et al. [9]	Natural language and structural similarity	Business process in XML	Medium
Montani et al. [22]	Structural similarity	Process log	Medium
Yan et al. [33]	Attribute similarity	BPMN notation	Medium
Niemann et al. [23]	Natural language and graph similarity	SAP reference model	Very Low
Dijkman et al. [8]	Behavioural, natural language and graph similarity	SAP reference model	Very Low
Zha et al. [36]	Behavioural similarity	Transition adjacency relation set	Very Low
Lu et al. [21]	Structural, Behavioural, and natural language	Business process constraint network (BPCN), and process variant repository (PVR)	Low

(continued)

Table 1. (*continued*)

Publication	Type similarity	Format of input data	Criteria match
Jung et al. [14]	Structural similarity	Non specified process model is converted to: weighted Complete Dependency Graph (wCDG)	Very Low
Dijkman et al. [6]	Natural language and graph similarity	SAP reference model	Low
Jung and Bae[15]	Behavioural similarity	Weighted complete dependency graphs,	Very Low
Huang et al. [12]	Graph similarity	Weighted complete dependency graphs,	Very Low

As shown in Table 1, most of the authors used more than one type of similarity techniques. None of the studies focused on RPAs, nor did they utilize RPA source codes or log information. This is confirmed by Schoknecht et al. [29] in their literature review. Most approaches would require transforming the RPA process into a specific input format in order to be usable. For example, converting RPA code into BPMN has already been proposed in some approaches: [10, 13, 28]. The transformation would then be less demanding than with other approaches. The least amount of effort for utilizing an existing method for finding similarity would be to use methods that utilize process logs [22, 37], or other studies that did not appear in the searched results [1, 2].

In Table 1, the criteria match column shows a range of values from very low to medium. These values indicate a match with the criteria presented earlier in this paper. None of the techniques in Table 1 would fulfil all of the criteria. The closest ones were the algorithms which used similar input data to RPAs such as process logs or XML, or which made use of the BPMN format because of its easy transformation from RPA code. Also, some algorithms were valued higher because of a natural language similarity, attribute similarity or other similarity approaches which would be useful in the RPA domain.

3   Description of Method

Our proposed method SiDiTeR (Similarity Discovering Techniques for RPA) uses natural language-based and graph similarity-based methods. The method is composed of three main parts. The first part is the decomposition of the RPA process/design. The second part of SiDiTeR focuses on natural language matching. The activities from RPA process are compared with activities in a provided dictionary feature (later referred as dictionary Δ), and this then produces an abstract (meta) process. The third part of SiDiTeR is the use of the longest common sequence (LCS) algorithm to find the longest sequences in the processes.

3.1 SiDiTeR

In the first part, SiDiTeR decomposes the source code of the RPA process, referred to then as the RPA design. From the design, we extract all of the activities with a name α. We preserve the order of activities α in the RPA design. Technically, we extract the activity names after the colon tag starting with < ui: from the XAML files. An example is < ui:ReadRange. We extract just the name ReadRange from the text. Thanks to the decomposition, we are able to have an RPA design activity list **A** for each process that we decompose this way and save to a list of all activities **A**.

SiDiTeR then creates a new activity list λ for each design. Then it searches through all activities α in the activity list **A** and looks for a match in the dictionary of identical activities Δ (see Table 2). If no match is found, it adds the activity to the new list λ with an original name. When a match is found between activity α and activity δ from the dictionary Δ, activity α is assigned a more abstract description (a meta-action name in Table 2) of activity δ that describes what the activity does. This results in a more abstract process i.e. meta process of the activity, which is stored in the newly created list λ. This results in a list of lists denoted as Λ. This process is visualized for an example in Fig. 2.

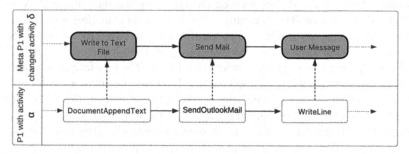

Fig. 2. Conversion to a meta process

The third part of SiDiTeR is a search for the longest common sequence for every meta process λ saved in Λ. The longest common sequence algorithm finds identical sequences in all newly made meta processes. The found sequences have to be equal to or longer than 3 activities in order to qualify for saving. The saved activities allow the user to effectively search for similar processes activities which can be then refactored. The user later has to make decisions if the component is the same and should be refactored into reusable components for another RPA process. An example of a found common sequence in two processes is shown in Fig. 3. For understanding this process better, a description of the code is written below. See Code 1.

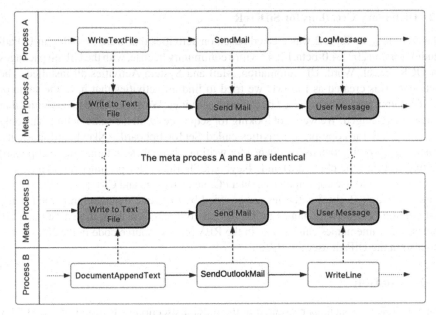

Fig. 3. Example of comparing meta processes

Code 1. Pseudocode of the SiDiTeR

```
list of activities A
list of lists Λ
for each `A in A:
  new list λ
  for each α in `A:
    if α match δ in Δ:
      λ add δ
    else:
     λ add α
    Λ add λ
function LongestCommonSequence(Λ):
  dictionary lcs
  for each pair of lists in Λ find identical sequence of
activities:
  if LEN(sequence) >= 3:
    lcs add sequence
  return lcs
```

3.2 Dictionary Creations for SiDiTeR

The dictionary in Table 2 was created based on activities that were available in UiPath Studio, version 2023.4.0-beta.12241 with a community license, with the UiPath packages for OCR, Excel, Word, Ui Automation, Mail and System Activities all installed. The dictionary was created as follows: we tried to find all activities that have the same or similar output but can be achieved by different activities. We only looked for activities that can be interchanged. We were not looking for sequences with the same output. The only exception was for copy-paste activities called SetToClipboard - NkeyboardShortcuts, which also work together as a sequence for writing. Using these actions, SetToClipboard (setting text into the clipboard) and NkeyboardShortcuts (for pasting), are identical to how a user would use copy and paste on a PC, i.e., Ctrl + C and Ctrl + V. In the second column of Table 2, the δ activities are grouped by the same meta-action named in the first column. The activity names in the second column come from the UiPath activity names. The same names can be seen in the RPA process source code in the XAML file and also in the UiPath user interface.

4 Evaluation

Our SiDiTeR approach, as presented in the previous section, was tested on a real RPA process made for UiPath. We programmed SiDiTeR in Python 3.11 to evaluate our approach. The repository with the sample processes is publicly available[1]. We evaluated the effectiveness of SiDiTeR on 156 UiPath process designs. Among the processes were 120 various sample processes, such as setting up an email account, calculator, robotic enterprise framework, executing commands in PowerShell and many others. The processes were in.xaml format and came from public repositories from GitHub or UiPath. In the dataset there were also 36 corporate automations from the banking industry which are not publicly available, and they are under a non-disclosure agreement. The corporate process comes from one banking company, and their process is used in the UiPath Robotic Enterprise Framework for building RPA processes. This is nicely presented in the results, where 15 files from 36 corporate process files have the longest common sequence of 163 same activities in the files. The second longest common sequence is 36 activities, and it comes from a different version of robotic enterprise framework files. The rest of the sequence is much shorter, and it would be important go through the activities manually and evaluate them. All of the results from SiDiTeR are presented in Table 3. In total, we were able to discover 655 matches among the tested xaml files.

At the outset, we proposed the following research aim:

To explore the possibility of finding similarities or identical parts in an RPA process for refactoring if many automations were developed by people who no longer work in a particular company, or if the development was outsourced.

This research paper demonstrates that it is possible to identify similar or identical parts in an RPA process. The results show that SiDiTeR can identify the same or similar activities across RPA processes and help the RPA developers or RPA maintenance team identify the activities which are candidates for refactoring.

---

[1] Available on Github: https://github.com/Scherifow/SiDiTar or Zenodo: https://zenodo.org/badge/latestdoi/644473852.

Table 2. Dictionary Δ

Meta action name	Activity name
Write in UI	NTypeInto, SetToClipboard - NKeyboardShortcuts, CVTypeIntoWithDescriptor
Write to Text File	WriteTextFile, WordAppendText, DocumentAppendText, AppendLine, DocumentReplaceText, WriteTextFile,NTypeInto
Write to Spreadsheet	WriteCSVFile, WriteCellX, AppendCsvFile, WriteRangeX, AutoFillX, ExportExcelToCsvX, InvokeVBAX,CopyPasteRangeX, AppendRangeX, AutoFitX, FindReplaceValueX, AppendRange, WriteCell, WriteRange, ExecuteMacroX, OutputDataTable, AddDataRow, UpdateRowItem, NTypeInto
Creation of Data Objects	BuildCollection < Object >, CreateList < Object >, BuildDataTable
Write to Data Objects	AppendItemToCollection < Object >, AppendItemToList < Object >, UpdateListItem < Object >, AddDataRow, UpdateRowItem
SAP login	Login, Logon,
OCR	GoogleCloudOCR, MicrosoftAzureComputerVisionOCR, CjkOCR, GoogleOCR, UiPathDocumentOCR, UiPathScreenOCR
Send Mail	SendMail, SendOutlookMail, SendMailX
Receive Mail	GetPOP3MailMessages, GetOutlookMailMessages, GetIMAPMailMessages
Save Mail	SaveMail, SaveOutlookMailMessage, SaveMailX
User Message	LogMessage, WriteLine
Get text	CVGetTextWithDescriptor, NGetText, GetOCRText
Click	CVClickWithDescriptor, Nclick, ClickOCRText
Hover	CVHoverWithDescriptor, Nhover, HoverOCRText
Highlight	CVHighlightWithDescirptor, Nhighlight
Extract DataTable	CvExtractDataTableWithDescriptor, NExtractData
Read File Text	DocumentReadText, WordTextRead, ReadTextFile
Save to clipboard	SetToClipboard, CopySelectedText
Loop	ForEach < Object >, InterruptibleWhile,InterruptibleDoWhile, ParallelForEach < Int32 >
Condition	If, IfElseIf, Switch < Int32 >

5 Discussion and Limitations

We have proposed a new method for discovering similarity in RPA processes (SiDiTeR). SiDiTeR uses an RPA design for the analysis of similar parts of different processes. This helps to refactor RPA code into reusable components more easily. The results show that SiDiTeR is able to find candidates among RPA processes for refactoring. As mentioned

Table 3. Results

Length of longest sequence	Number of found values
3	481
4	125
5	30
6	2
9	1
36	1
163	15

in the introduction, this is one of the solutions for overcoming an RPA maintenance trap, as the whole portfolio of RPA bots will then be more easily governed [16]. To find out which part of an RPA process should be refactored into components, process similarity techniques can be used.

In the field of process similarity, there has been a decline in the number of new works published [29]. The use of process equivalence and process similarity techniques in the field of RPA can be a new spark for more research and publications in the field. With a higher number of RPA automations, there will be a higher demand for making the automations sustainable and avoiding the RPA maintenance trap. As seen from the related works, no technique has addressed this topic yet. Thus, this could be an impulse for using process similarity in another practical application.

We are aware of certain limitations that our approach currently has. One of the concerns is that SiDiTeR works only with UiPath designs, and the dictionary is made for UiPath activities. This limitation concerning UiPath designs is easily addressable, at least partially, and it would be enough to decompose the activity names from the source code of another platform. The limitation concerning the dictionary is more complicated, as partial knowledge of the platform is needed to create a similar dictionary. It is likely that the size of the dictionary will be different for different platforms. In certain cases, such as writing vs copying and pasting text, these activities can be adopted one to one for other platforms. When creating the dictionary for our study, only activities that had identical or similar resulting actions were used. The dictionary could be extended to include sequences where the output of the activities is also identical, but the result achieved is made up of multiple actions such as: clicking in the UI vs using a keyboard shortcut; or, for example, using the UI instead of using the API. Experienced programmers are likely to use the most efficient path, but for junior development or citizen development, inefficient sequences are likely to occur [17, 27].

SiDiTeR can also raise questions about why we use process similarity techniques for processes instead of techniques from the computer science field, even though RPA is software. This is a justified question because there are already techniques for code refactoring. For example, a systematic literature review from 2020 [4], analyzed 41 techniques concerning automatic software refactoring. But we focused more on process similarity due to the fact that RPA process (code) can also be analyzed as a process.

RPA as a process is more understandable to non-technical users, citizen developers and process owners. The understanding by stakeholders of a process can by crucial for the additional validation of refactoring of the correct part of a process. The main advantage of SiDiTeR techniques is that they can be used on the source code of RPA or also on the process log to analyze the RPA as a process.

Another limitation may be the accuracy of SiDiTeR, where in some cases the activities are not identical but will still be included, even though they are different processes i.e. false positives. Accuracy could be increased by using parameters and incorporating attribute similarity into SiDiTeR. This approach would then be even more efficient for users who will evaluate the results. There is an opportunity for extending this research further, for the purpose of identifying the right candidates for refactoring among RPA processes more precisely.

6 Conclusion

Finding similarity in the RPA domain is very useful, because it can be used for refactoring. The refactoring of RPA processes will be one of the crucial components for future RPA governance, since the same parts of RPA code can be refactored into components and shared across a portfolio of RPA bots. We have presented a new approach for detecting identical or similar parts in RPA processes called SiDiTeR. SiDiTeR is designed with RPAs in mind, and can easily read RPA code or process logs with nested activities and handle extra activities in processes. It can also deal with different activities with the same output, which is crucial for complex refactoring. Our approach was tested on 156 RPA processes. The longest match we discovered was with 163 activities across 15 processes and 655 matches among RPA processes. These results challenge future researchers to find ways to identify parts of RPA which could be more precise, and thus allow for a more convenient search method for suitable components for refactorization.

Acknowledgment. This research was made possible thanks to the Technical University of Liberec and the SGS grant number: SGS-2023–1328. This research was conducted with the help of Pointee.

References

1. van der Aalst, W.M.P., de Medeiros, A.K.A., Weijters, A.J.M.M.: Process equivalence: comparing two process models based on observed behavior. In: Dustdar, S., Fiadeiro, J.L., Sheth, A.P. (eds.) Business Process Management. BPM 2006. Lecture Notes in Computer Science, vol. 4102. Springer, Heidelberg (2006). https://doi.org/10.1007/11841760_10
2. Alves de Medeiros, A.K. et al.: Quantifying process equivalence based on observed behavior. Data Knowl. Eng. **64**(1), 55–74 (2008). https://doi.org/10.1016/j.datak.2007.06.010
3. Amiri, M., Koupaee, M.: Data-driven business process similarity. IET Softw. **11**(6), 309–318 (2017). https://doi.org/10.1049/iet-sen.2016.0256
4. Baqais, A.A.B., Alshayeb, M.: Automatic software refactoring: a systematic literature review. Softw. Qual. J. **28**(2), 459–502 (2020). https://doi.org/10.1007/s11219-019-09477-y
5. Cao, B., et al.: Querying similar process models based on the Hungarian algorithm. IEEE Trans. Serv. Comput. **10**(1), 121–135 (2017). https://doi.org/10.1109/TSC.2016.2597143

6. Dijkman, R., Dumas, M., García-Bañuelos, L.: Graph matching algorithms for business process model similarity search. In: Dayal, U., Eder, J., Koehler, J., Reijers, H.A. (eds.) Business Process Management. BPM 2009. Lecture Notes in Computer Science, vol. 5701. Springer, Heidelberg (2009). https://doi.org/10.1007/978-3-642-03848-8_5

7. Dijkman, R., et al.: Managing large collections of business process models—current techniques and challenges. Comput. Ind. **63**(2), 91–97 (2012). https://doi.org/10.1016/j.compind.2011.12.003

8. Dijkman, R., et al.: Similarity of business process models: metrics and evaluation. Inf. Syst. **36**(2), 498–516 (2011). https://doi.org/10.1016/j.is.2010.09.006

9. Figueroa, C., et al.: Improving business process retrieval using categorization and multimodal search. Knowl.-Based Syst. **110**, 49–59 (2016). https://doi.org/10.1016/j.knosys.2016.07.014

10. Flechsig, C., et al.: Robotic process automation in purchasing and supply management: a multiple case study on potentials, barriers, and implementation. J. Purchas. Suppl. Manage. **28**(1), 100718 (2022). https://doi.org/10.1016/j.pursup.2021.100718

11. Garcia, M.T., et al.: BPMN-Sim: a multilevel structural similarity technique for BPMN process models. Inf. Syst. **116**, 102211 (2023). https://doi.org/10.1016/j.is.2023.102211

12. Huang, K., et al.: An algorithm for calculating process similarity to cluster open-source process designs. In: Jin, H., et al. (eds.) Grid and Cooperative Computing GCC 2004 Workshops, Proceedings, pp. 107–114 (2004)

13. Hüller, L., et al.: Ark automate — an open-source platform for robotic process automation. In: Proceedings of the Demonstration and Resources Track, Best BPM Dissertation Award, and Doctoral Consortium at BPM 2021 co-located with the 19th International Conference on Business Process Management, CEUR Workshop Proceedings, Rome, Italy (2021)

14. Jung, J., et al.: Hierarchical clustering of business process models. Int. J. Innov. Comput. Inf. Control **5**(12A), 4501–4511 (2009)

15. Jung, J., Bae, J.: Workflow clustering method based on process similarity. In: Gavrilova, M., et al. (eds.) Computational Science and Its Applications - ICCSA 2006, PT, vol. 2, pp. 379–389 (2006)

16. Kedziora, D., Penttinen, E.: Governance models for robotic process automation: the case of Nordea Bank. J. Inf. Technol. Teach. Cases **11**(1), 20–29 (2021). https://doi.org/10.1177/2043886920937022

17. Klimkeit, D., Reihlen, M.: No longer second-class citizens: Redefining organizational identity as a response to digitalization in accounting shared services. J. Prof. Organ. **9**(1), 115–138 (2022). https://doi.org/10.1093/jpo/joac003

18. Klinkmuller, C., Weber, I.: Analyzing control flow information to improve the effectiveness of process model matching techniques. Decis. Support Syst. **100**, 6–14 (2017). https://doi.org/10.1016/j.dss.2017.06.002

19. Liu, C., et al.: Measuring similarity for data-aware business processes. IEEE Trans. Autom. Sci. Eng. **19**(2), 1070–1082 (2022). https://doi.org/10.1109/TASE.2021.3049772

20. Liu, C., et al.: Towards comprehensive support for business process behavior similarity measure. IEICE Trans. Inf. Syst. **E102D**(3), 588–597 (2019). https://doi.org/10.1587/transinf.2018EDP7127

21. Lu, R., et al.: On managing business processes variants. Data Knowl. Eng. **68**(7), 642–664 (2009). https://doi.org/10.1016/j.datak.2009.02.009

22. Montani, S., et al.: A knowledge-intensive approach to process similarity calculation. Expert Syst. Appl. **42**(9), 4207–4215 (2015). https://doi.org/10.1016/j.eswa.2015.01.027

23. Niemann, M., et al.: Comparison and retrieval of process models using related cluster pairs. Comput. Ind. **63**(2), 168–180 (2012). https://doi.org/10.1016/j.compind.2011.11.002

24. Niu, F., et al.: Measuring business process behavioral similarity based on token log profile. IEEE Trans. Serv. Comput. **15**(6), 3344–3357 (2022). https://doi.org/10.1109/TSC.2021.3104898

25. Pei, J., et al.: Efficient transition adjacency relation computation for process model similarity. IEEE Trans. Serv. Comput. **15**(3), 1295–1308 (2022). https://doi.org/10.1109/TSC.2020.298 4605

26. Průcha, P.: Aspect optimalization of robotic process automation. In: ICPM 2021 Doctoral Consortium and Demo Track 2021, CEUR Workshop Proceedings, Eindhoven, The Netherlands (2021)

27. Průcha, P., Skrbek, J.: API as method for improving robotic process automation. In: Marrella, A., et al. Business Process Management: Blockchain, Robotic Process Automation, and Central and Eastern Europe Forum. BPM 2022. Lecture Notes in Business Information Processing, vol. 459. Springer, Cham (2022). https://doi.org/10.1007/978-3-031-16168-1_17

28. Rybinski, F., Schüler, S.: Process discovery analysis for generating RPA flowcharts. In: Marrella, A., et al. Business Process Management: Blockchain, Robotic Process Automation, and Central and Eastern Europe Forum. BPM 2022. Lecture Notes in Business Information Processing, vol. 459. Springer, Cham (2022). https://doi.org/10.1007/978-3-031-16168-1_15

29. Schoknecht, A., et al.: Similarity of business process models-a state-of-the-art analysis. ACM Comput. Surv. **50**, 4 (2017). https://doi.org/10.1145/3092694

30. Sohail, A., et al.: An intelligent graph edit distance-based approach for finding business process similarities. CMC-Comput. Mater. Continua **69**(3), 3603–3618 (2021). https://doi.org/10.32604/cmc.2021.017795

31. Thaler, T., Schoknecht, A., Fettke, P., Oberweis, A., Laue, R.: A comparative analysis of business process model similarity measures. In: Dumas, M., Fantinato, M. (eds.) Business Process Management Workshops. BPM 2016. Lecture Notes in Business Information Processing, vol. 281. Springer, Cham (2017). https://doi.org/10.1007/978-3-319-58457-7_23

32. Valero, V.: Strong behavioral similarities in timed-arc Petri nets. Appl. Math. Comput. **333**, 401–415 (2018). https://doi.org/10.1016/j.amc.2018.03.073

33. Yan, Z., et al.: Fast business process similarity search. Distrib. Parallel Databases **30**(2), 105–144 (2012). https://doi.org/10.1007/s10619-012-7089-z

34. Ye, Z., et al.: Synthesis of contracted graph for planar nonfractionated simple-jointed kinematic chain based on similarity information. Mech. Mach. Theory **181** (2023). https://doi.org/10.1016/j.mechmachtheory.2023.105227

35. Zeng, Q., et al.: A novel approach for business process similarity measure based on role relation network mining. IEEE Access **8**, 60918–60928 (2020). https://doi.org/10.1109/ACCESS.2020.2983114

36. Zha, H., et al.: A workflow net similarity measure based on transition adjacency relations. Comput. Ind. **61**(5), 463–471 (2010). https://doi.org/10.1016/j.compind.2010.01.001

37. Zhou, C., et al.: A comprehensive process similarity measure based on models and logs. IEEE Access **7**, 69257–69273 (2019). https://doi.org/10.1109/ACCESS.2018.2885819

38. Robotic Process Automation (RPA) Software Reviews 2023 | Gartner Peer Insights. https://www.gartner.com/reviews/market/robotic-process-automation-software. Accessed 21 Apr 2023

What Are You Gazing At? An Approach to Use Eye-Tracking for Robotic Process Automation

A. Martínez-Rojas[1(✉)], H. A. Reijers[2], A. Jiménez-Ramírez[1],
and J. G. Enríquez[1]

[1] Departamento de Lenguajes y Sistemas Informáticos, Escuela Técnica Superior de Ingeniería Informática, Avenida Reina Mercedes, s/n, 41012 Sevilla, Spain
{amrojas,ajramirez,jgenriquez}@us.es
[2] Department of Information and Computing Sciences, Utrecht University, Princetonplein 5, 3584 CC Utrecht, The Netherlands
h.a.reijers@uu.nl

Abstract. User Interface (UI) logs are crucial in capturing and analyzing user behavior, enabling a comprehensive understanding of business processes and eventual process automation with Robotic Process Automation (RPA). However, extracting meaningful insights from UI logs becomes challenging, especially when dealing with complex and information-dense graphical user interfaces. This paper presents a novel approach that leverages eye-tracking technology to address this challenge. The proposed solution incorporates gaze fixation (i.e., where the user pays attention to the user interfaces) into the UI log, which is then used to filter irrelevant information from it. Two gaze-based filtering methods are presented and evaluated using synthetic and real-life screenshots. Preliminary results demonstrate that the method effectively reduces the irrelevant UI elements by an average of 76% while keeping meaningful information on the screen.

Keywords: Robotic Process Automation · User Interface · Eye-tracking · Relevance detection

1 Introduction

In recent years, Robotic Process Automation (RPA) has witnessed unprecedented growth in the industrial landscape, which has fueled research in this discipline and the rising of new subfields, such as Task Mining [1,16] or Robotic Process Mining [9]. It consists of automatically discovering repetitive routines that can be automated using RPA. This is done by analyzing interactions between users and software applications during the performance of tasks in a business process.

This publication is part of the projects PID2019-105455GB-C31 and PID2022-137646OB-C31, funded by MCIN/ AEI/10.13039/501100011033/ and by the "European Union"; the FPU scholarship program, granted by the Spanish Ministry of Education and Vocational Training (FPU20/05984) and its mobility grants (EST23/00732).

J. Köpke et al. (Eds.): BPM 2023, LNBIP 491, pp. 120–134, 2023.
https://doi.org/10.1007/978-3-031-43433-4_8

Although Task Mining has been successfully applied in a wide range of businesses, some challenges still need to be faced. In general, the amount of data extracted when monitoring user behaviors can be unwieldy [9]. This problem is exacerbated when considering the content of the user interfaces (UIs) [10]. In these cases, feature extractors are used to extract such content for the screenshots which typically generate a large number of features, i.e., UI elements. Therefore, analyzing RPA processes using all the features on the screen becomes challenging since irrelevant information can cause meaningful insights to be lost (cf. Fig. 1) [10].

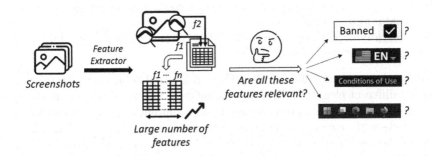

Fig. 1. Motivation problem.

To tackle this issue, we aim to use eye-tracking systems to record the user's gaze and discern relevant from irrelevant information on the screenshots. The current paper extends this previous work [11], which presents a method to integrate a new type of event, the "attention point change", within the UI log, the so-called User Behaviour Log (UB log). Here, we extend this concept with additional information, i.e., gaze fixation and dispersion, which are needed to define attention areas instead of attention points.

Moreover, the present work offers:

- A conceptualization of gaze events, screenshots, and UI components. Therefore, fixation, dispersion, and other properties can be studied over the components in the screenshot.
- A method to incorporate information from these gaze events into the UI log. Thereby, the resulting UB log can filter the content of the screenshots.
- Two filtering methods to discard all parts of the screen that do not receive attention. These methods are evaluated using both synthetic and real-life screenshots.

The rest of the paper is organized as follows. Section 2 introduces the background concepts. Section 3 describes the related work. Section 4 explains the proposed approach. Section 5 evaluates the approach and analyzes the results obtained. Section 6 discusses the results and the applicability of the approach. Finally, Sect. 7 summarizes the proposal and exposes future work.

2   Background

2.1   Eye-Tracking

An eye-tracker is a device that uses infrared light sources and/or cameras to monitor and analyze eye movements. Thus, eye-tracking is a technique used to measure and analyze eye movements during visual processing [4,6]. This technique includes the projection of the eyes' sight on the screen to disclose where the human is gazing at. The points on the screen at which the human gazes can be grouped into different kinds of eye movements. The most representative are:

- **Fixations**: periods of stable gaze that occur when the eyes are relatively still, and visual information is processed.
- **Saccades**: rapid, jerky eye movements that occur when the eyes shift from one fixation point to another.

By tracking these eye movements, researchers gain insights into how attention is focused on different elements. This links to medical and psychological research which emphasizes that accurate perception and comprehension depend on fixating on objects with our eyes, commonly referred to as attention [15]. Therefore, fixation can be used to determine which elements on the screen the user is paying attention to.

2.2   UI Components Detection

Object detection is a computer vision technique that aims to identify and locate specific objects within an image or video. Object detection can be applied to detect and recognize graphical user interface (GUI) components displayed within a software application in the context of screenshots. To detect GUI components, traditional computer vision, deep learning, or both can be used [20].

There is a vast number of terms to conceptualize GUI components, e.g., GUI elements [20], GUI components, GUI groups, or GUI containers [12]. Nonetheless, the literature in this field has focused on two main concepts [2]: *UI elements* detection which aims to detect GUI components at one level (i.e., without considering hierarchy between them) [2,12,20], and *UI groups* detection, that, on the contrary, considers hierarchies. These terms can be defined as follows [3,12]:

- **UI element**: atomic graphical element with predefined functionality on which the user action is executed, e.g., buttons, text boxes, dropdowns, checkboxes, or sliders.
- **UI group**: an entity that groups UI elements and defines spatial display properties of their members, forming a hierarchical structure. The *UI group* that represents the root of the hierarchical structure receives the special name of *System*.

From hereafter, *UI component* will be considered as an abstract term to indistinctly refer to every type of component that appears on the screen, either *UI elements* or *UI groups*.

2.3 UI Log and Feature Extraction

UI logs are frequently employed to record user interactions with an information system during process execution. This phase of recording is referred to as behavior monitoring (cf. Fig. 2, step 1). Several proposals [8,9] use them to capture events. They may differ, but share commonalities that can be summarized by the following definition: A UI Log extends the concept of an event log by including additional attributes that provide details about the corresponding UI interactions. These attributes can be categorized into three groups: process attributes (e.g., timestamp, event ID, case ID), application attributes (e.g., app name, screen capture, screen size), and user input attributes (e.g., type of action, click coordinates, keystrokes).

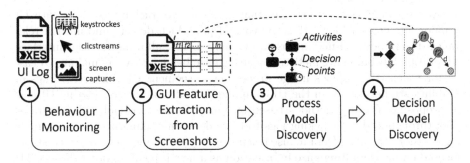

Fig. 2. Example of application of feature extraction in RPA (extracted from [10]).

Additionally, information from screenshots can be incorporated as structured information in the UI logs through GUI feature extractors [10]. GUI feature extractors are software components that extract specific features from screenshots. In this paper, we refer to GUI feature extractor as feature extractor for the sake of simplicity. These features extractors are applied to each event in the UI log (i.e., to its corresponding screenshot), resulting in a new attribute being added to the log for each extracted feature (cf. Fig. 2, step 2) as described in [10]. As a result, screen features are associated with log events, enabling further analysis of this data. This analysis facilitates process discovery (cf. Fig. 2, step 3) and decision models (cf. Fig. 2, step 4) that explain the decisions made in the process. Consequently, it uncovers the decisions made during this process for potential automation, taking into account screen information. However, including a large number of features in the log can overwhelm decision discovery models with irrelevant information. The motivation of this work is to reduce the log's feature set to address this issue.

3 Related Work

In recent years, eye-tracking technology has gained interest in various domains of process science, such as data visualization, understandability, and decision-making. In [18], eye-tracking is used to improve data visualizations for decision-making, while [7] analyzes eye-tracking data using process mining to identify reading patterns during comprehension tasks.

Studies focusing on process model representation and eye-tracking have also been conducted. Specifically, [3] investigates user engagement with different artifacts in a hybrid process model representation. Moreover, [19] examines visual routines and gaze patterns involved in process model comprehension. In addition, [14] proposes task-specific visual cues to enhance process model understanding, and [15] explores performance variations in different model comprehension tasks.

Although these studies are not directly related to the goal of this paper, they share the common approach of utilizing eye-tracking fixations as a measure of user attention and connecting the concept of relevant information to the areas that capture attention.

Alternatively, there are studies that combine eye-tracking fixations with UI components or event logs. It is the case of [5], which detects information relevance during decision-making using eye movements, allowing for user interface adaptation by combining metrics capturing gaze behavior. Along making UI easier for users, [17] employs image processing to segment UI video recordings in usability studies to obtain the UI components, but without considering UI logs or proposing filtering methods based on image frames. Instead, [13] integrates click-through and eye-tracking logs to mine decision-making processes, extracting decision strategies from mouse tracing and eye-tracking data. In contrast, we focused on incorporating gaze information as a new type of event to UI logs [11] to obtain a UB log. In addition, this UB log definition is extended to record user attention fixations. Based on this log, two alternatives to filter process-relevant information are presented: gaze pre-filtering and post-filtering.

4   Approach

This section describes our approach to filter irrelevant information from the screenshots based on the gaze events. Firstly, Sect. 4.1 includes (1) the formalization of the entities needed to structure the information from the gaze interaction and the screenshots and (2) the way of combining the information of gaze and UI events to build a UB log. Secondly, Sect. 4.2 defines the different filtering methods that can be applied over this UB log.

4.1   Understanding Interaction and Representation

Eye-trackers are commonly used in research to study fixation patterns. They relate fixations to areas that are most relevant to the human [6]. Consequently, in this study, gaze events are defined in such a way that they store information related to these types of patterns:

Definition 1. *A Gaze Event is defined as an event type that stores the following attributes from a gaze interaction between the human and the information system:*

- *The type property indicates whether the event is a fixation or a saccade.*
- *The x and y values represent the x and y coordinates related to the user's gaze on the screen.*
- *The #events property records how many gaze points (i.e., fixation or saccade) the human has looked at during the gaze event.*
- *The start and end properties record the beginning and ending timestamps of the event.*
- *The duration property records the length of the gaze event in milliseconds (i.e., the difference between end and start).*
- *The dispersion property records the scatter in the gaze position during the event.*

Example 1. An example of a gaze event would be: *"type"*: fixation, *"x"*: 819.916, *"y"*: 584.666, *"#events"*: 3, *"start (ms)"*: "35,162.196" *"end (ms)"*: "36,727.022" *"duration"*: "1,564.826", *"dispersion"*: "15.994"

GazeEvent: Type=Fixation

Timestamp	EventType	CoordX, CoordY	TextInput	NameApp	Screenshot	Fixation X, Fixation Y	#events	...	dispersion
10001	Mouse	123,32		Mail client	001.png				
10002	GazeFixation			CRM	002.Png	374,23	16	...	12.56
10003	GazeFixation			CRM	002.Png	355,212	23	...	10.73
10004	GazeFixation			CRM	002.Png	135,633	11	...	11.53
10005	Mouse	234,367	"283633J"	CRM	002.Png				
10006	GazeFixation			Mail client	003.Png	151,636	31	...	15.15
10007	GazeFixation			Mail client	003.Png	351,661	25	...	09.51
⋮	⋮	⋮	⋮	⋮	⋮	⋮			⋮

Fig. 3. User Behaviour Log with Gaze Fixation Events added.

To incorporate the gaze event into the UB log, they are introduced by extending the UI log in the same way as in [11]. For gaze events, the *eventType* is "GazeFixation" and the *timestamp* is the *start* time of the gaze event. The rest of the UI log properties remain the same in the UB log, but it includes the properties of the gaze events (cf. Fig. 3). Although both types of events (i.e., saccade and fixation) are recorded by the eye-tracker, the current paper ignores the saccade ones since they lack meaningful information regarding the user's attention.

These fixation events place the user's attention at certain coordinates on the screen. To match the gaze coordinates with the UI components within the screen, it is necessary to formalize and structure the information on the screen. For this purpose, the *Screen Object Model* is defined to support the main types of UI component detection (cf. Sect. 2.2) and its hierarchy on the screen.

Definition 2. *A Screen Object Model (SOM) is a data structure that represents the information on a screenshot. Its design is inspired by the Document Object Model (DOM) which is the standard for representing information on a website, but SOM simplifies it by focusing just on the static and observable parts of the screen. Therefore, it is useful for any kind of system, besides websites. A SOM consists of the following elements:*

- *Filename: name of the source file it represents, e.g., "screenshot001.jpg".*
- *ScreenShape: list with the height, width, and the number of color channels stored in the source image that it represents, e.g., $[800, 1422, 3]$.*
- *System: The top UICompo on the screen. An UICompo is a UI element or a group of them that appears on the screen defined as a tree data structure with the following properties:*
  - *id: as a unique identifier.*
  - *bbox: a tuple of (x1,x2,y1,y2) representing the bounding box of the UI component in the screen.*
  - *type: to indicate whether it is a "UI Group" (i.e., it has children), a "UI Element" (i.e., it is a tree leaf), or "System" (i.e., it is the root of the tree).*
  - *children: list of UICompo in the current UICompo. In the case of UI elements, it is empty.*
  - *label: it is a tuple <class, text> that contains the content of the current element. The class refers to the class of the UI element (e.g., button, input text, etc.), and the text refers to the textual content of such a component. This textual information depends on the class of the element, e.g., for a button, the text contains the title of the button.*

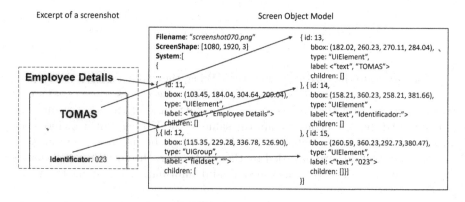

Fig. 4. Example of a Screen Object Model related to a crop of a screenshot.

The Screen Object Model follows a tree structure in which the root is always the UICompo with the *System* type (cf. Fig. 4). UI Compos detected with one-level component detection algorithms can be stored as a list of children components of the *System*. Hierarchical component detection algorithms can fully exploit the tree structure.

4.2   Gaze Filtering

The current proposal uses a gaze filtering technique to eliminate irrelevant UI components from screenshots by analyzing the user's attention. With this aim, two alternative algorithms are proposed for gaze filtering: **pre**-filtering and **post**-filtering (cf. Fig. 5).

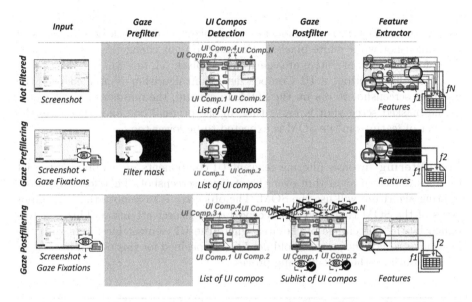

Fig. 5. Phases included in the pre-filtering and post-filtering approaches.

Pre-filtering involves applying a gaze-based mask to the related screenshot. Therefore, the screenshot will be retaining the regions with fixation while discarding the rest. The aim of this technique is to narrow the focus to these specific areas for further analysis. Algorithm 1 outlines the pre-filtering process.

The pre-filtering algorithm starts by loading the gaze fixation events that have occurred over the input screenshot to the input screenshot (cf. line 1). Then a binary mask is initialized, represented as a matrix of 0s (i.e., black pixels), with the same dimensions as the screenshot (cf. line 2). The algorithm iterates through each fixation event and draws an attention area as a circle of 1s on the mask (i.e., white pixels). This attention area has the fixation point as the center, and the dispersion determining its radius (cf. lines 3-5). Finally, the

---
Algorithm 1. Gaze Prefiltering
---
input: UB log *ubLog*
input/output: Screenshot *screenshot*
  1: GazeEvent[] *gEvents* ← Load gaze events of *screenshot* from *ubLog*
  2: Mask *mask* ← Initialize a mask as a zero matrix of the same size as *screenshot*
  3: **for** *gazeEvent* in *gEvents* **do**
  4:     *mask* ← drawCircle(*gazeEvent*.fixation, *gazeEvent*.dispersion)   ▷ Set as 1 a
    circle of pixels centered at fixation point with dispersion as the radius
  5: **end for**
  6: *screenshot* ← Apply *mask* over *screenshot*, setting non-attentive areas as black
---

mask is applied to the screenshot, i.e., black pixels in the mask become black in the screenshot while white pixels in the mask do not change in the screenshot. Therefore, the non-attentive areas are turned black (cf. line 6).

After this filtering, a UI Compos detection algorithm is applied over the modified screenshot to obtain the SOM. In general, the number of UI elements in the masked screenshot is lower than in the original one and, therefore, the number of features in the SOM is expected to be lower too.

Post-filtering involves filtering relevant UI components *after* applying the UI Compos detection algorithm over the original screenshot. Therefore, the post-filtering algorithm will use the SOM. The objective is to remove the UI components of the SOM that have not received enough user attention. This is determined by the percentage of the area where the UI element intersects with the fixation area. For this, a threshold need to be specified for this percentage. Algorithm 2 outlines the post-filtering process:

Algorithm 2. Gaze Postfiltering
---
input: Screenshot *screenshot*, UB log *ubLog*, Threshold *thresh*
input/output: SOM *som*
  1: GazeEvent[] *gEvents* ← Load gaze events from *ubLog* for *screenshot*
  2: Set *relevantsCompos* ← {}
  3: **for** *gazeEvent* in *gEvents* **do**
  4:     **for** *uiCompo* in *uiCompos* **do**
  5:         **if** intersection_area(*uiCompo*.bbox, *gazeEvent*.attention_area) > *thresh*
    then
  6:             *relevantCompos* $\xleftarrow{+}$ *uiCompo*
  7:         **end if**
  8:     **end for**
  9: **end for**
 10: removeOpposite(*som*, *relevantCompos*)

Similar to the pre-filtering algorithm, the post-filtering algorithm starts by loading the gaze events associated with the specific screenshot from the UB log (cf. line 1) and initializes the set of relevant UICompos as empty (cf. line 2). Then the algorithm iterates through each gaze event and checks for intersections between the bounding box of the UI components and the attention area of the gaze event (cf. lines 3-5). This area is calculated in the same way that in the pre-filtering, a circle centered in *gazeEvent.fixation* point with *gazeEvent.dispersion* as radius. If the intersection area exceeds a threshold, the corresponding UI component's *XPath* is included in the SOM, i.e., all the UICompos in the path from the System UICompo until the current UICompo are marked as relevant (cf. line 6). Finally, the irrelevant UICompos (i.e., the opposite set of the relevant ones) are removed from the SOM.

In general, the number of UI components that overlap with gaze event areas is lower than the complete set of UI components and, therefore, the number of features in the SOM is expected to be lower too after applying the post-filtering algorithm.

Pre-filtering focuses on reducing the attentional focus within the screenshot itself, while post-filtering operates on the obtained SOM including UI components based on their intersection with gaze events. Both pre- and post-filtering techniques offer different approaches to gaze filtering. The choice between these techniques should be based on the specific problem context and the goals of the analysis.

By implementing gaze filtering techniques, we can reduce the number of features extracted from the UI elements, improving computational efficiency and potentially enhancing the accuracy of the subsequent analysis applied to the log enriched with these filtered features.

5 Evaluation

This section evaluates the effectiveness of our two approaches in reducing the number of features while retaining the relevant UI elements. Section 5.1 describes the set-up that underlies our evaluation, and Sect. 5.2 provides the data analysis and summarizes the results of the evaluation.

5.1 Set-Up

In this evaluation, a single subject, i.e., a 25-year-old male wearing contact lenses with astigmatism, assessed the proposed gaze filtering methods. The evaluation involved real and mockup screenshots. Pre and post-filtering algorithms were applied to study (1) their performance in reducing components while keeping relevant components, and (2) the efficiency of component detection. Each screenshot had one or two key UI components referred to as *relevants*, representing the areas the user gazed at, which were crucial for performing actions on the screenshots.

The filtering algorithms were applied to 5 different groups of screenshots (P). Each group contained 10 screenshots of the same type[1]:

---

[1] The set of problems is available at: https://doi.org/10.5281/zenodo.8009445.

- P1: mockup screenshot representing the visualization of email content. The relevant component of this screenshot was an attachment file.
- P2: mockup screenshot representing the detailed view of a user in a Customer Relationship Management (CRM) platform. The relevant component of this screenshot was the status of a checkbox (i.e., checked or unchecked).
- P3: real screenshot representing the visualization of email content (Gmail). The relevant component was the same as P1.
- P4: real screenshot representing the detailed view of a user in a CRM platform (Odoo). The relevant component was the same as P2.
- P5: real screenshot representing a split screen containing two applications, i.e., a pdf viewer, and the detailed view of an employee in a customized human resources management system. In this case, there are two key components, one per application. The action performed involved comparing the first and last name in the PDF with those in the employee's detail view, making them the key components for these screenshots.

To analyze the results obtained from the pre and post-filtering algorithm, they were compared with the case where the capture had not been filtered. The following measures were studied for each problem:

- $\#UIElem$: counts the UI elements in the resulting SOM.
- $UIDet(s)$ SOM detection time, in seconds. It measures the time it takes to detect the SOM from a screenshot. For this, an adaptation of the tool [20] is used.
- $\%Reduction$: determines the percentage reduction in the number of UI elements compared to the number of UI elements in the unfiltered case.
- $\%Relevant?$: determines the percentage of relevant UI elements detected after the filtering.

Note that, the $\#UIElem$ in the $Non-filtered$ case represents the number of UI elements that exist in the screenshot since no filter has been applied.

5.2 Results

The evaluation was executed in an *ASUS Rog Strix G513IC* laptop, with an *AMD Ryzen 7 4000 series* CPU, 16 GB of RAM, and Windows 11 as the operating system. Results obtained for each metric are presented in Table 1.

It is noteworthy that since the Non-filtered case does not apply any filter, the $\%Reduction$ is 0% and the $\%Relev?$ is 100%. Likewise, as the Post-filtering case applies the UI Component detection algorithm before the filtering, the $UIDet(s)$ is the same as in the non-filtered case.

The evaluation indicates that this approach successfully minimizes the number of UI elements while preserving the relevant ones in both algorithms. This reduction leads to a decrease in the feature set, which allows for improved computational efficiency and potential accuracy enhancements in subsequent tasks involving these features. Analyzing the results in-depth, it can be seen that in

Table 1. Evaluation results

	Non-Filtered				Pre-filtering				Post-filtering			
P	#UIElem	%Reduct	%Relev?	UIDet(s)	#UIElem	%Reduct	%Relev?	UIDet(s)	#UIElem	%Reduct	%Relev?	UIDet(s)
P1	18	0	100	5,5	2,2	87,8	60	4,3	3,5	80,6	100	5,5
P2	19	0	100	5,9	8,1	57,4	80	4,9	10,0	47,4	100	5,9
P3	118	0	100	7,7	7,0	94,1	70	4,8	11,6	90,2	100	7,7
P4	106	0	100	12,9	18,1	82,9	80	5,3	23,8	77,6	100	12,9
P5	165	0	100	15,0	37,5	77,3	100	7,5	58,2	64,7	100	15,0

terms of reduction, pre-filtering remove a higher number of UI elements than post-filtering. In contrast, in terms of relevance, post-filtering consistently outperforms pre-filtering. It maximizes the preservation of relevant UI elements. This can be explained by the fact that post-filtering recovers the UI element even if part of its area lies outside the attention area while the pre-filtering masks part of the UI element.

Related to the time, pre-filtering demonstrates a notable decrease in UI component detection times, contributing to time efficiency in the UI component detection phase. This reduction can be explained by the fact that masking nonrelevant areas in the screenshot will reduce the computation time of the used detection algorithms.

In summary, this evaluation highlights the advantages of both pre and postfiltering techniques. Pre-filtering offers gains in UI element reduction and time efficiency, while post-filtering ensures the retention of relevant UI components.

6 Discussion

The choice of an eye-tracking tool and its algorithm for determining fixations has an impact on the evaluation results. In our study, the iMotions solution is used[2], which follows its own criteria for fixation and saccade extraction. This biases the calculation of the attention areas by relying on fixation and dispersion and, therefore, may affect the results. Moreover, the performance obtained by the eyetracking tool is intimately related to the filtering performance. In this case, the subject recorded in eye-tracking sessions was a person that had astigmatism. This means that the calibration of the eye-tracking tool was not optimal. A better calibration could provide better filtering results.

Regarding the UI component detection phase, UIED solution [20] is employed for the detection. However, this method introduces errors, particularly while detecting nested components. To mitigate these issues and ensure accurate results, a manual review was conducted to refine the UI component detection process. Additionally, the use of UIED relates to how the pre-filtering is implemented. It sets to black the non-attended areas. This removing method is assumed to be compatible with the UIED component detection. However, if a different detection technique is used after pre-filtering, adjustments might be necessary to ensure compatibility.

---

[2] iMotions Eye Tracking System: https://imotions.com/eye-tracking/#Solution.

During the post-filtering process, a threshold of 0.25 is utilized to determine the percentage of the intersection area between gaze fixations and the UI component, which is considered relevant. This specific value has been chosen based on favorable results obtained through trial and error. It is crucial to explore and compare multiple threshold values to understand their impact on the approach. This exploration is important both in terms of reducing the data and optimizing performance for future use of the filtered features. For instance, to study how this parameter influences the identification of decision models applied to post-filtered data.

7   Conclusions and Future Work

In RPA, grasping relevant features from user interfaces is paramount. These features help to develop decision rules to discover why users interact with specific UI components or to build scripts to automate user actions. Nonetheless, an overabundance of irrelevant information in the UI can generate numerous noisy features, thereby impairing the effectiveness of decision discovery algorithms. Recent research highlights the efficacy of eye-tracking data —particularly metrics related to fixations— to identify pertinent information displayed on the screen. Thus, this study combines eye-tracking data and a UI structured representation (i.e., the so-called Screen Object Model) and creates filtering algorithms to reduce the number of features to extract from the user interfaces, while retaining information about relevant UI components.

The evaluation demonstrates the effectiveness of our approach in reducing the number of features while preserving the relevant UI components. By significantly decreasing the feature set, our method optimizes computational efforts and potentially improves the accuracy of subsequent tasks involving these features.

Although this study has provided good insights into the understanding of relevant features in RPA, there are several promising directions for future investigation: (1) study how relevance is affected by metrics other than fixation (e.g., saccade records or the number of times a user look at a UI element); (2) determine the optimal algorithm parameters configuration (e.g., thresholds, the radius of gaze dispersion, etc.) depending on the context in which the analysis is made; (3) study the importance of the fixated data for candidates for automation in RPA contexts, and (4) evaluate other conventional eye-tracker tools (e.g., webcam-based) which would allow their use in a wider range of cases and budgets.

References

1. van der Aalst, W.M.P.: Process mining: data science in action, 2nd edn. Springer, Heidelberg (2016). https://doi.org/10.1007/978-3-662-49851-4
2. Abb, L., Rehse, J.R.: A reference data model for process-related user interaction logs. In: Di Ciccio, C., Dijkman, R., del Rio Ortega, A., Rinderle-Ma, S. (eds.) Business Process Management. BPM 2022. LNCS, vol. 13420, pp. 57–74. Springer, Cham (2022). https://doi.org/10.1007/978-3-031-16103-2_7

3. Abbad Andaloussi, A., Slaats, T., Burattin, A., Hildebrandt, T.T., Weber, B.: Evaluating the understandability of hybrid process model representations using eye tracking: first insights. In: Daniel, F., Sheng, Q.Z., Motahari, H. (eds.) BPM 2018. LNBIP, vol. 342, pp. 475–481. Springer, Cham (2019). https://doi.org/10. 1007/978-3-030-11641-5_37

4. Brunyé, T.T., Drew, T., Weaver, D.L., Elmore, J.G.: A review of eye tracking for understanding and improving diagnostic interpretation. Cogn. Res. Princ. Implic. **4**(1), 1–16 (2019). https://doi.org/10.1186/s41235-019-0159-2

5. Feit, A.M., Vordemann, L., Park, S., Berube, C., Hilliges, O.: Detecting relevance during decision-making from eye movements for UI adaptation. In: ACM Symposium on Eye Tracking Research and Applications, pp. 1–11 (2020)

6. Gwizdka, J.: Characterizing relevance with eye-tracking measures. In: Proceedings of the 5th Information Interaction in Context Symposium, pp. 58–67 (2014)

7. Ioannou, C., Nurdiani, I., Burattin, A., Weber, B.: Mining reading patterns from eye-tracking data: method and demonstration. Softw. Syst. Model. **19**, 345–369 (2020)

8. Jimenez-Ramirez, A., Reijers, H.A., Barba, I., Del Valle, C.: A method to improve the early stages of the robotic process automation lifecycle. In: Giorgini, P., Weber, B. (eds.) CAiSE 2019. LNCS, vol. 11483, pp. 446–461. Springer, Cham (2019). https://doi.org/10.1007/978-3-030-21290-2_28

9. Leno, V., Polyvyanyy, A., Dumas, M., La Rosa, M., Maggi, F.M.: Robotic process mining: vision and challenges. Busin. Inf. Syst. Eng. **63**, 301–314 (2021)

10. Martínez-Rojas, A., Jiménez-Ramírez, A., Enríquez, J.G., Reijers, H.A.: Analyzing variable human actions for robotic process automation. In: Di Ciccio, C., Dijkman, R., del Rio Ortega, A., Rinderle-Ma, S. (eds.) Business Process Management. BPM 2022. LNCS, vol. 13420, pp. 75–90. Springer, Cham (2022). https://doi.org/10. 1007/978-3-031-16103-2_8

11. Martínez-Rojas., A., Jiménez-Ramírez., A., Enríquez., J.G., Lizcano-Casas., D.: Incorporating the user attention in user interface logs. In: Proceedings of the 18th International Conference on Web Information Systems and Technologies - WEBIST, pp. 415–421. INSTICC, SciTePress (2022). https://doi.org/10.5220/0011568000003318

12. Moran, K., Bernal-Cárdenas, C., Curcio, M., Bonett, R., Poshyvanyk, D.: Machine learning-based prototyping of graphical user interfaces for mobile apps. IEEE Trans. Software Eng. **46**(2), 196–221 (2018)

13. Petrusel, R.: Integrating click-through and eye-tracking logs for decision-making process mining. Inform. Econ. **18**(1), 56 (2014)

14. Petrusel, R., Mendling, J., Reijers, H.A.: Task-specific visual cues for improving process model understanding. Inf. Softw. Technol. **79**, 63–78 (2016)

15. Petrusel, R., Mendling, J., Reijers, H.A.: How visual cognition influences process model comprehension. Decis. Support Syst. **96**, 1–16 (2017)

16. Reinkemeyer., L.: Process mining in action. principles, use cases and outlook, 1st Edn. Springer, Cham (2020). https://doi.org/10.1007/978-3-030-40172-6

17. Simko, J., Vrba, J.: Screen recording segmentation to scenes for eye-tracking analysis. Multimedia Tools Appl. **78**, 2401–2425 (2019)

18. Weber, B., Gulden, J., Burattin, A.: Designing visual decision making support with the help of eye-tracking. In: RADAR+ EMISA@ CAiSE, pp. 47–54 (2017)

19. Winter, M., Neumann, H., Pryss, R., Probst, T., Reichert, M.: Defining gaze patterns for process model literacy-exploring visual routines in process models with diverse mappings. Expert Syst. Appl. **213**, 119217 (2023)
20. Xie, M., Feng, S., Xing, Z., Chen, J., Chen, C.: UIED: a hybrid tool for GUI element detection. In: Proceedings of the 28th ACM Joint Meeting on European Software Engineering Conference and Symposium on the Foundations of Software Engineering, pp. 1655–1659 (2020)

Automating Computer Software Validation in Regulated Industries with Robotic Process Automation

Nourhan Elsayed[1], Luka Abb[1(✉)], Heike Sander[2], and Jana-Rebecca Rehse[1]

[1] University of Mannheim, Mannheim, Germany
{elsayed,luka.abb,rehse}@uni-mannheim.de
[2] DHC Business Solutions GmbH & Co. KG, Saarbrücken, Germany
heike.sander@dhc-vision.com

Abstract. Computerized System Validation (CSV) aims to ensure that software systems in regulated environments, such as the pharmaceutical industry, adhere to strict regulatory requirements. It plays a crucial role in guaranteeing the safety of products or services by verifying that computerized systems operate according to the specified guidelines. However, CSV is a complex and resource-intensive process that poses challenges for small and medium-sized enterprises in particular, making it challenging to be implemented effectively. In this paper, we investigate the potential of Robotic Process Automation (RPA) as a solution to partially automate CSV and address these challenges. We present an ongoing project where we apply RPA to CSV and discuss its effectiveness in reducing the time and effort associated with manual CSV activities. We also describe the challenges that we encountered during the implementation of RPA in the context of CSV. Our research highlights the possibility of extending the application of RPA beyond simple data entry and verification tasks, allowing for the automation of entire complex processes.

Keywords: computerized system validation · regulated industries · robotic process automation

1 Introduction

The use of computerized systems in regulated environments such as the pharmaceutical industry is subject to strict regulations, which aim to ensure that the systems adhere to specifications that guarantee the safety of people or products used by people. Good practice (GxP) quality guidelines and regulations exist to guide the use of software systems in regulated environments and require quality implementation, use, maintenance, and documentation of software systems. Companies in regulated industries must provide documented proof that their software systems used comply with the requirements for verification purposes. This documented proof is the result of the Computerized System Validation (CSV) process, which consists of steps, tests, and checks that ensure the software is designed and executed as intended [9,18].

J. Köpke et al. (Eds.): BPM 2023, LNBIP 491, pp. 135–148, 2023.
https://doi.org/10.1007/978-3-031-43433-4_9

Ensuring the compliance of GxP-relevant software and processes with these regulations is costly and resource-intensive. CSV must be executed throughout the entire system development lifecycle: Systems must be validated before deployment and remain in a validated state afterwards. Configuration changes must follow formalized change processes so that the mapping of processes, technical components, and validation tests remains traceable. Consequently, validating GxP-relevant software and processes is complex and often very expensive, which makes it particularly challenging for small and medium-sized companies with limited budgets. These organizations may face difficulties in effectively implementing CSV, risking a compliance violation [14]. Automating parts of CSV, in particular the labor-intensive testing of validation requirements [10], would therefore be a significant benefit for companies in regulated industries.

One promising automation technology that has gained significant attention in the context of business process management is Robotic Process Automation (RPA), which involves the use of software robots to automate routine and repetitive tasks traditionally performed by humans [2]. These bots are capable of mimicking human interactions with computer systems, enabling them to execute tasks in a fast and consistent manner. RPA has been successfully applied in various domains, including finance, healthcare, and customer service [16].

In the CSV context, RPA has the potential to (partially) automate validation, enabling organizations to reduce the required time and effort. RPA can enhance traceability in the validation process by automatically documenting each step performed by the bots, ensuring compliance with regulatory requirements. However, although RPA offers considerable advantages, its implementation in the context of CSV is much more challenging compared to the automation of simple tasks. In a typical task automation scenario, such as data entry, data validation, or document processing [4], it is sufficient that a software robot executes the task as intended by the user. In contrast, in CSV, it is also essential to verify that the *outcome* of the automated task execution is correct and to ensure that the task execution is documented according to regulations.

In this paper, we present a semi-automated approach to CSV, which we develop in an ongoing research project. This approach leverages RPA, along with ideas from the robotic process mining framework [15]. Instead of relying on explicitly configured bots, we record task executions and utilize user interaction logs (UI logs) as input for automation. In realizing the project, we encountered several challenges, including regulatory concerns, process aspects, and implementation details. The paper is intended as an experience report that offers insights into the complexities involved in automating intricate business processes like CSV. It thus adds to a relatively small body of existing literature exploring the applications of RPA beyond isolated task automation [7,11,13].

The rest of this paper is structured as follows: we introduce CSV and some key terms in Sect. 2. In Sect. 3, we describe our approach and its expected benefits compared to manual validation. The challenges we encountered are discussed in Sect. 4. In Sect. 5, we outline the next steps we plan to take in our project, before concluding the paper in Sect. 6.

2   Computerized System Validation

GxP Regulations. "Good Practices" (GxP) regulations are the international pharmaceutical requirements and regulations that need to be followed in regulated industries, such as pharmaceutical or biotechnology industries. The "x" in GxP can stand for different applications, such as Good Quality Practice (GQP), Good Manufacturing Practice (GMP), or Good Laboratory Practice (GLP). These regulations are put in place to guide companies' operations to ensure "patient safety, product quality, and data integrity" [9,18]. Computerized system validation (CSV) is the process of ensuring that a software system or application meets specified requirements, performs as intended in a given context, and is in compliance with relevant GxP standards and regulations [6]. With the increasing reliance on software applications in regulated industries, their computerized systems are also subject to GxP regulations. Thus, robust software validation practices are crucial for these industries.

CSV Project Stages. Compliance with GxP regulations can be realized by following a systematic, good-practice-based system lifecycle, from conceptualizing the system and identifying its requirements to system planning, configuration, and verification, to system operations, and finally system retirement. Most CSV activities take place during system planning, configuration, and verification. A widely recognized standard framework for validating computerized systems, which many practitioners follow, is the so-called V model [18], shown in Fig. 1. It provides a structured approach to ensure that the software meets requirements and functions effectively. The V model consists of a series of stages, including planning, specification, configuration, verification, and reporting or documentation. These steps ensure thorough testing, risk assessment, and compliance with regulatory requirements [8,12].

According to the V model, the validation process begins with *planning*, where validation experts define the validation objectives, scope, and necessary resources [8]. This stage lays the groundwork for subsequent steps and ensures alignment with relevant regulations and business processes. The next step is *specification*, where software requirements are specified based on applicable regulations and standards. Specifications include user, functional, and design specifications [8, 12]. Specifications may differ from system to system according to the required system functionality [18]. After specification, system *configuration* and software *coding* take place according to the defined standards [18].

Next is *verification*, which involves conducting a comprehensive review of the software design and code to ensure it aligns with the defined requirements and that the specifications are met [8]. Verification includes different testing activities [18,21]. Test execution entails creating and executing test cases to assess the software's functionality, performance, and compliance. Once the phase is complete, the results are analyzed to identify any potential issues or deviations. This analysis helps in understanding the underlying causes of defects or failures encountered during testing and facilitates improvements to the software [12].

Multiple documents are involved in this phase, such as the *test plan* which depicts the test schedule and approach for testing the system. All documents,

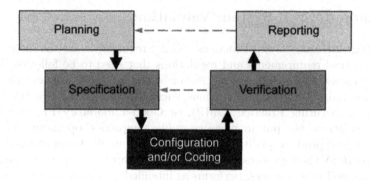

Fig. 1. Simplified V Model of a CSV process

from requirements to *Test scripts*, are traceable, meaning they include references between the requirement and the tests necessary to confirm the implementation of the requirement. Test scripts include the instructions for the test steps that need to be executed and the acceptance criteria that need to be satisfied for a successful test case execution. Different test cases can be created to test the same requirements, i.e., using different test data or testing different results. *Test scripts* are detailed instructions to execute test steps. Once the testing is conducted, test result reporting should take place. Test reports include successful and failed tests and act as documented evidence of system status [18].

The last validation phase is *reporting/documentation*, which already begins during validation and all documents must be released before the system is released for productive use [18]. GxP regulations require a systematic approach to handling documentation in CSV projects. Different documents used in the validation project, such as validation reports and testing documents, must have a structured format and must include accurate, complete, and traceable information. All activities and outcomes, including test plans, test cases, test results, and any deviations or issues faced, need to be documented. This documentation serves as a reference for future validation activities, audits, and compliance purposes. It ensures transparency, reproducibility, and traceability throughout the validation process [8].

Once the system is validated, the organization must ensure compliance with GxP regulations throughout its operational lifespan. There should always be up-to-date documentation that the system is in a validated state. Additionally, any system changes should also be validated and documented according to standards to ensure regulatory compliance. Thus, the V model phases will have to be repeated throughout the system's lifespan and when any system changes are introduced [18].

3   Valid-AI-te: Towards Semi-automated CSV with RPA

In this section, we present an ongoing research project in which we aim to semi-automate Computerized System Validation with Robotic Process Automation.

Specifically, our goal is to automate the testing and documentation of the test cases (verification and reporting stages of the V model) that verify if a system is in a validated state. Planning, specification and configuration of requirements would remain as manual activities.

The corresponding software tool, *valid-AI-te*, is currently under internal development at our project partner, a software vendor that specializes in regulated industries. It will be tested with pilot customers from these industries once development is complete.

3.1   Approach Description and Implications for CSV

The central component of Valid-AI-te is an automated testing tool that uses test documentation as initial input. This test documentation contains instructions at the user interface (UI) level, explaining how a validation user should navigate through a system to execute a test case. It also includes acceptance criteria for each step. The automation approach requires the validation user to execute the test case once while using a recording component to capture all actions in a UI log. This log follows the data model introduced in [3] and is used as input to an RPA script that executes the recorded steps. Once the initial recording is complete, the user can *parametrize* it to specify the exact execution of the test case or to adapt it to further test cases. There are two types of parametrization: One allows the user to determine values for text fields, such as naming a newly created object, while the other enables actions on specific elements, like selecting a document from a list. An overview of the testing procedure is shown in Fig. 2.

The fundamental capabilities of this automation tool are similar to those found in many other tools for automated software testing [5,10,19,20]. However, they only cover part of the CSV process, namely the execution of test scripts according to test documentation. In order to also address the verification of acceptance criteria, as well as outcome documentation, reporting, and traceability requirements, Valid-AI-te includes several additional components.

In order to automatically verify the required acceptance criteria at predetermined stages of the test case, valid-AI-te allows the validation user to configure *acceptance checks*. These checks follow predefined templates and are implemented in three ways: 1. As database queries to verify conditions in the application's backend, such as the existence and/or status of an object. This approach considers the backend database as a source of truth, which deviates from the traditional visual validation in the frontend. 2. The status of frontend objects (such as disabled/enabled buttons, displayed lists, etc.) which are not manifest in any database, can be determined as well. This needs to be adapted to the system that is to be validated. 3. Lastly, screenshots can be used as supplementary sources of truth and to verify the existence/appearance of an object in the frontend for further verification.

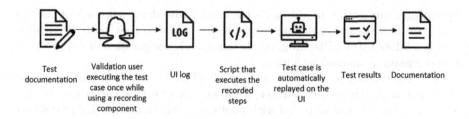

| Test documentation | Validation user executing the test case once while using a recording component | UI log | Script that executes the recorded steps | Test case is automatically replayed on the UI | Test results | Documentation |

Fig. 2. Overview of the approach steps

Parametrization, acceptance checks, and screenshot points are inputs that can be configured prior to the execution of automated tests. The output of the tests is produced by the *documentation and reporting* component of valid-AI-te, which creates an outcome report for each executed test case. The report documents the actions undertaken by the software robots and includes a list of the performed acceptance checks with their parameters, binary outcomes (pass or fail), additional information if a check fails, as well as screenshots and the option for the user to confirm the binary outcome. If all acceptance checks pass, the report serves as proof that the validation requirement(s) covered by the test case have been fulfilled.

Individual test cases can also be combined into a comprehensive *scenario*, where the success of individual test cases is the prerequisite for a successful continuation of the test scenario. Thus, in addition to the acceptance checks and user-led acceptance via screenshots, there is a third performance review that is based on the successful execution of a combination of dependent test cases, which makes it possible to test higher-order scenarios as well. The essential traceability requirement of CSV is then secured through the correlation of test cases, acceptance criteria, and results, as well as the documentation of inter-dependencies within test scenarios.

The automation approach thus requires the user to be able to record, config-ure and manage test cases, and to monitor and approve the test outcomes. An example mockup of a corresponding user interface is shown in Fig. 3.

3.2   Expected Benefits

In the initial validation stage, i.e., when a GxP-relevant system is first intro-duced, valid-AI-te and manual CSV follow the same approach: each test case needs to be executed once by a qualified validation user, according to a test document that has been derived from the User Requirements Specification. The main advantage of valid-AI-te lies in the ability to automatically execute the test case recordings whenever the system undergoes a configuration change, and to automatically produce the documentation needed to be compliant with reg-ulatory requirements. The validation users only need to configure the test cases (i.e., supply parameters and define acceptance checks), which is significantly less time-intensive than executing and documenting every test case manually. In

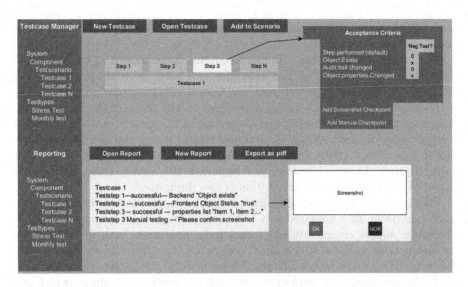

Fig. 3. A mockup showing test case configuration and reporting. Test cases and scenarios can be combined from previously recorded logs, as shown in the upper part. The lower part shows a report of a performed test.

many cases, it is sufficient to only configure an automated test case once and re-use it for future validation processes, further reducing the required manual effort. For example, the *create new document* test case always has the same acceptance checks and can be executed with the same document name in both the initial validation and the validation of a software change that affects the software components required to run this process. Of course, this only applies if the process itself or the UI does not change; in these cases, a new recording or a new parametrization would be required.

Valid-AI-te also makes it possible to run automated test routines at regular intervals. Manual CSV, due to its high cost, is relevant after every critical system change, where the criticality/risk is assessed after every change. Therefore, most companies only allow scheduled, intended system changes, which is not feasible for example in systems that are not hosted on-premise and run the risk of not performing critical updates due to a lack of validation resources. Valid-AI-te reduces the testing cost, thus making it possible to perform a complete system check after every update or change and to omit the risk assessment. However, in the case of process or UI changes, the UI logs would have to be re-recorded. The testing scope can be increased and the threshold for testing lowered.

A further scenario where cheap automated testing and documentation are beneficial are process-neutral system changes that occur in the backend of the application but should not be visible in the frontend. Usually, if a (process-neutral) change in the system occurs, the criticality analysis and definition of testing scope must be repeated in order to identify affected processes and testing needs. With automated testing, all processes can be (re-)tested without additional resource investment.

The traditional validation approach requires users trained in the processes, the system, and validation. Test scenarios can cover multiple processes that require more than one expert to perform. Coordinating training and availability of those users can be challenging, depending on the scope of the implementation project. This issue is mitigated, as testing of the system can be automated independently of user training. Even if user testing is required by company policies, in case of availability issues, Valid-AI-te can play the role of a second user during testing. As documentation is automated, the need to train users in the formalities of validation documentation is also diminished.

Finally, even when users are trained, executing test cases and verifying acceptance criteria in an automated way is less error-prone than manual validation. The latter almost completely relies on human judgment, both in interpreting the instructions in the test document and in deciding whether what is visually shown in the user interface fulfills the acceptance criteria. In the automated approach, the potential error sources are reduced to the initial recordings (which can easily be verified by replaying them) and the parametrization (for example, a user might specify to select an object which does not exist). In previous validation projects, we found that during testing, more than 50 % of errors are either due to wrong usage of the system or due to misunderstandings or unclear test scripts/test plans, The occurrence of these errors in individual system users ranges from 0 to 80% dependent on the experience of the users with the system. Both usage errors, as well as test script/test plan errors, can be avoided by automated testing.

4   Challenges of Automating CSV with RPA

In this chapter, we describe the challenges that we faced when automating the CSV process with RPA and suggest possible strategies to overcome them.

4.1   Regulatory-Level Challenges

Complying with Legislation
Description: If the valid-AI-te solution is to be deployed in production and to replace the established manual validation approach, it must be ensured that it is able to meet all regulatory requirements pertaining to CSV. However, there are currently no specific norms, guidelines, or governmental regulations that specifically consider an AI-based validation approach. At the very least, the applicability of this new method requires approval from regulatory authorities, which would need to be secured first.

Strategies to overcome the challenge: A key aspect to overcoming this challenge is to demonstrate that the results generated by our approach, i.e., the output documents, are of a quality and reliability comparable to those obtained through conventional manual inspection procedures. With regulatory agencies now increasingly focusing on the use of AI, we believe that the acceptance of valid-AI-te within the framework of innovative solution approaches in the CSV field is likely, but not guaranteed.

Ensuring Privacy
Description: Another challenge relates to privacy concerns, particularly when sensitive information, such as usernames and passwords, is entered by users during test cases. It is crucial to ensure that this sensitive data is not recorded in the logs, while still allowing the execution of essential tasks such as login.

Strategies to overcome the challenge: To address this issue, we will implement an event abstraction [22] approach that identifies activities performed in (manually determined) privacy-sensitive areas of the system, abstracting them to generic activities manageable by the automation component. For instance, in the login scenario, the bot would read out the abstracted action from the UI log and then use its dedicated credentials to log in. Expanding the scope of the project, such as through continuous monitoring of system interactions and validation of associated tasks (see Sect. 5), would introduce additional privacy challenges. To address these challenges, we may need to leverage techniques from privacy-preserving process mining [17].

Validation of Valid-AI-te
Description: Software used in the regulated environment must be validated. Introducing an automated RPA tool, while facilitating the validation of the main system, requires tool validation as well, which might introduce a similar amount of work that is saved.

Strategies to overcome the challenge: By giving the user information about the test execution in the reports and allowing the user to make the ultimate passed/failed decision as well as giving the user the information on which decisions are based, validation of the tool can be omitted. However, validation of the recorded processes is required, as those are equivalent to validation documentation (test case descriptions). This validation must be performed by process/validation experts prior to the use of the automated tests and can be offered by the software developer of both Valid-AI-te as well as the software to be tested. Therefore, Valid-AI-te is a tailored in-house solution for the specific software it is being configured for.

4.2   Process-Level Challenges

Tailored Validation Processes
Description: For every validation project, target GxP processes for the company are derived from available reference models of GxP processes to serve as the basis for validating the computerized system. This makes validation requirements and activities specific to the company's case. Accordingly, different automated validation execution scenarios should be developed tailored to the company's specific requirements and target processes.

Strategies to overcome the challenge: While there is the need for individualization of processes, these new processes can be recorded for the individual case as well, and thus, valid-AI-te can be customized to individual companies. In addition, as logs can be recombined, only the new process components must be re-recorded and often standard components might be reused. This allows the creation and editing of test cases tailored to the individual companies' needs.

Incomplete Coverage of Possible Validation Scenarios

Description: (GxP) Processes in computerized systems have a degree of freedom in how can they be (correctly) executed. For example, it may be possible to open the same menu with two different buttons, or, as described above, the order in which tasks are performed may vary. Ideally, all execution scenarios should be identified and validated for checking compliance with regulations (although this does not always happen in manual validation). Identifying possible acceptable behavior for executing the process and differentiating it from deviating behavior is challenging in both the classical and automated validation approach.

Strategies to Overcome the Challenge: This challenge is classically addressed by the mitigation measure "training", however, there is never a 100% safety that users are not using the system differently, especially since if the varieties do not affect the successful performance of the tests. We can address this challenge by providing the ability to identify and record different process execution scenarios to incorporate different acceptable execution behavior as well as a flexible strategy for test case design that makes it possible to incorporate different positive and negative test scenarios.

Complex Processes

Description: (GxP) Processes in a computerized system differ in their complexity. Some consist of a small number of execution steps, whereas others involve a larger number of steps that leads to complex execution. When recording the execution of such complex processes, the generated UI logs will be very long, and therefore difficult to comprehend and work with. In particular, they might be very challenging for validation users to parametrize correctly.

Strategies to Overcome the challenge: We have developed several auxiliary functionalities in order to facilitate the handling of long UI logs. For instance, the validation user can during recording specify that the last action he recorded should eventually be parametrized. The corresponding row will then include a visual indicator. To supplement the UI log, the user can also take screenshots during recording and thereby document the UI state at any point. Connected processes can be identified manually during recording or be connected later into test scenarios, thus changes in parameters in one aspect of these processes demands a change of the parameter, wherever else it is reused and thus allows the user to identify the need for updated parametrization.

Unstructured Acceptance Criteria

Description: Handling acceptance checks to be automatically checked after test case steps is challenging because (1) in the manual validation, they are not represented in a structured format, (2) there can be an arbitrary number of acceptance checks after each step in a test case, and (3) it is not always clear how the different outcomes of each acceptance check could be handled (for example, if the test case should continue even though an acceptance check has failed). In addition, certain acceptance criteria likely cannot be verified in an automated way and must be manually checked by the validation user.

Strategies to Overcome the Challenge: We have developed a structured format (parametrized templates) for the types of acceptance checks that need to be performed in our project. We currently rely on users to configure these acceptance checks, although in the future it may be possible to automate this by deriving them from the textual description in the test documents. We also plan to devise an outcome-handling strategy for each acceptance check template. For example, we would allow the test case execution to continue if a user has not received a mail notification about the start of an object workflow, since this deviation has no effect on the following steps. However, we would abort the test case if the object in the workflow is of the wrong object type. To accommodate hybrid automated/user-led acceptance checking, interactive reports will be created in which the users can manually confirm if a test passed or failed.

4.3   System-Level Challenges

Recording and Automation in Complex User Interfaces
Description: Creating recording and automation scripts that can work on various types of potentially dynamic UI elements is challenging. A key issue is the construction of appropriate identifiers for these elements because they are not directly available from the user interface and need to be interpretable to validation users for the purpose of parametrization. As an example, consider a recording in which a validation user creates a new document, then selects it from a table and initiates a workflow. Afterward, he wants to parametrize this test case to be executed with a different document. From a user perspective, the parameter to be supplied is the name of the document. However, the system does allow multiple documents of the same name to exist, which are differentiated by a document ID that the user cannot know before the document has been created. In this scenario, the parameter supplied by the user cannot directly be used as an identifier to select the correct UI element when executing the test case.

Strategies to overcome the challenge: We experimented with various out-of-the-box solutions for user interface recording, but none of these were able to address all the issues we encountered. Instead, we developed a recording approach that is tailored to the system in which we are implementing valid-AI-te. It relies on several indicator attributes in the UI log to uniquely identify elements in a way that is comprehensible to users. In the scenario sketched above, for instance, we record an attribute that indicates that the original UI element that represents the document is part of a dynamic table. We then select all elements in that table that match the user-supplied document name and use a decision rule to select one - specifically, we have decided that the choice that is most likely to be correct is the document that has been most recently created and therefore has the highest document ID.

Compatibility with Different Software Versions
Description: Computerized systems to be validated can have different versions and can run in different environments given the differences in companies' IT infrastructure. Additionally, there are multiple different computerized systems used in regulated environments and these systems are subject to validation. For this reason, managing the automation of computerized software validation across different versions and environments can be challenging.

Strategies to overcome the challenge: This can be overcome using existing validation documentation and information from previous validation projects from our partner company (check project description). This can reveal information about the types, and complexity of GxP-relevant systems to ensure building and automation solution that is compatible with different system configurations.

System Reaction Time
Description: Systems require a certain time to react, therefore it is possible that while the process works, it appears faulty because the system is too slow to react and the script tries to execute on something that is not loaded yet. Also, database updates after the creation of new objects take time (OOM minutes).

Strategies to overcome the challenge: This could be solved by ensuring/choosing appropriate waiting times in the approach. For instance, we can measure the execution/waiting time of the user and derive the waiting times between automated task execution from that.

5   Future Directions

Continuous Validation. Until now, we have focused on recording user interactions to test and validate computerized systems at specific points in time, for example, after a scheduled system change. However, it would also be possible to record any user interactions with the system and to continuously validate that the tasks that users perform to comply with regulatory requirements. To do so, we plan to augment the valid-AI-te solution with a *task recognition* component that recognizes user tasks while they are executed and can automatically perform basic acceptance checks. We plan to derive this recognition capability from the existing UI logs that were created in the initial validation. One option would be to apply supervised machine learning techniques to train a model to classify tasks based on the features recorded in the UI log. Another would be to apply task mining techniques to derive simple decision rules that the recognition component can apply (for instance, when the *New Document* button is clicked, it should recognize this as the start of the *Create document* task). This extension of the project scope brings with it first and foremost numerous technical challenges, but also additional regulatory concerns, since the constant monitoring of user activities could be considered an intrusion of privacy.

Conformance Checking. In the current scope of the project, we validate the results of the test cases, but not if they are executed and recorded correctly, or if the configured acceptance checks are appropriate. Ensuring this remains the responsibility of the validation users. In the future, we plan to also apply conformance checking techniques to the UI logs to ensure that the execution captured in them actually matches the test cases as they are described in the test case documents. In the absence of imperative process models, one possibility would be to apply an approach that automatically extracts declarative process constraints from the textual descriptions in the test case documents [1] and then verify that the recorded UI log does not violate these constraints.

Applications in IT Support. The recording component of valid-AI-te can be used to record arbitrary sequences of user interactions in the test system. This functionality can be leveraged in IT support by asking users to create a recording of the exact steps that led to a system issue. The resulting UI log is significantly more detailed and informative than a typical system error log and would make it easier for support agents to reproduce the steps the user has taken and narrow down the defect. We also plan to explore this application at some point in our research project.

6   Conclusion

In this paper, we have presented a research project in which we partially automate Computer System Validation with Robotic Process Automation. We have described the automation approach that we developed in this project and outlined the expected improvements over the classical validation process, as well as several regulatory and process-related challenges that we encountered.

Although RPA is an established technology when it comes to simple task automation, there has been little research on how well it can be applied to entire complex business processes. Through our project report, we offer insights into how this can be approached, what challenges one might encounter, and how these can potentially be solved. We believe that these insights are of interest to researchers and practitioners alike and will facilitate the broader adoption and application of Robotic Process Automation.

In terms of future work, we plan to explore additional applications of the components that we have developed in this project, particularly in continuous, real-time validation. Since there are currently very few UI logs publicly available, we also plan to at some point publish a dataset that contains the UI logs that we have recorded for valid-AI-te, along with the corresponding test documentation.

References

1. van der Aa, H., Di Ciccio, C., Leopold, H., Reijers, H.A.: Extracting declarative process models from natural language. In: Giorgini, P., Weber, B. (eds.) CAiSE 2019. LNCS, vol. 11483, pp. 365–382. Springer, Cham (2019). https://doi.org/10.1007/978-3-030-21290-2_23

2. Aalst, W.M.P., Bichler, M., Heinzl, A.: Robotic process automation. Bus. Inform. Syst. Eng. **60**(4), 269–272 (2018)
3. Abb, L., Rehse, J.R.: A reference data model for process-related user interaction logs. In: Business Process Management: 20th International Conference, BPM 2022, Münster, Germany, 11–16 September 2022, Proceedings, pp. 57–74. Springer (2022). https://doi.org/10.1007/978-3-031-16103-2_7
4. Aguirre, S., Rodriguez, A.: Automation of a business process using robotic process automation (RPA): a case study. In: Figueroa-García, J.C., López-Santana, E.R., Villa-Ramírez, J.L., Ferro-Escobar, R. (eds.) WEA 2017. CCIS, vol. 742, pp. 65–71. Springer, Cham (2017). https://doi.org/10.1007/978-3-319-66963-2_7
5. Alegroth, E., Nass, M., Olsson, H.H.: Jautomate: a tool for system-and acceptance-test automation. In: 2013 IEEE Sixth International Conference on Software Testing, Verification and Validation, pp. 439–446. IEEE (2013)
6. Altaie, A.M., Alsarraj, R.G., Al-Bayati, A.H.: Verification and validation of a software: a review of the literature. Iraqi J. Comput. Inform. **46**(1), 40–47 (2020)
7. Asquith, A., Horsman, G.: Let the robots do it! - taking a look at robotic process automation and its potential application in digital forensics. Forensic Sci. Inter. Reports **1**, 100007 (2019)
8. Beizer, B.: Software testing techniques, New York (1990)
9. Crumpler, S., et al.: General principles of software validation. Final Guidance Industry FDA Staff Version **2**, 1–47 (2002)
10. Fewster, M., Graham, D.: Software test automation. Addison-Wesley Reading (1999)
11. Huang, F., Vasarhelyi, M.A.: Applying robotic process automation (RPA) in auditing: a framework. Int. J. Account. Inform. Syst. **35**, 100433 (2019)
12. IEC: Iec 62304: 2006-medical device software-software life cycle processes (2006)
13. Jiménez-Ramírez, A.: Humans, processes and robots: a journey to hyperautomation. In: González Enríquez, J., Debois, S., Fettke, P., Plebani, P., van de Weerd, I., Weber, I. (eds.) BPM 2021. LNBIP, vol. 428, pp. 3–6. Springer, Cham (2021). https://doi.org/10.1007/978-3-030-85867-4_1
14. Kamalrudin, M., Sidek, S.: A review on software requirements validation and consistency management. Int. J. Softw. Eng. Appli. **9**(10), 39–58 (2015)
15. Leno, V., Polyvyanyy, A., Dumas, M., La Rosa, M., Maggi, F.: Robotic process mining: vision and challenges. Bus. Inform. Syst. Eng. **63** (2021)
16. Madakam, S., Holmukhe, R.M., Jaiswal, D.K.: The future digital work force: robotic process automation (RPA). JISTEM-J. Inform. Syst. Technol. Manag. **16** (2019)
17. Mannhardt, F., Koschmider, A., Baracaldo, N., Weidlich, M., Michael, J.: Privacy-preserving process mining. Bus. Inform. Syst. Eng. **61**(5), 595–614 (2019)
18. Shields, S.: GAMP 5: a risk-based approach to compliant GxP computerized systems. In: ISPE (2013)
19. Trudova, A., Dolezel, M., Buchalcevova, A.: Artificial intelligence in software test automation: A systematic literature review. In: International Conference on Evaluation of Novel Approaches to Software Engineering (2020)
20. Umar, M.A., Zhanfang, C.: A study of automated software testing: automation tools and frameworks. Int. J. Comput. Sci. Eng. (IJCSE) **6**, 217–225 (2019)
21. Wallace, D.R., Fujii, R.U.: Software verification and validation: an overview. IEEE Softw. **6**(3), 10–17 (1989)
22. van Zelst, S., Mannhardt, F., de Leoni, M., Koschmider, A.: Event abstraction in process mining: literature review and taxonomy. Granular Comput. **6**(3), 719–736 (2021)

Migrating from RPA to Backend Automation: An Exploratory Study

Andre Strothmann$^{(\boxtimes)}$ and Matthias Schulte

viadee Unternehmensberatung AG, Münster, Germany
{Andre.Strothmann,Matthias.Schulte}@viadee.de

Abstract. Robotic Process Automation (RPA) is often considered a short-term solution that bridges technology gaps temporarily. However, it is no longer the choice of automation technology when backend automation becomes readily available. We collect insights from thirteen interviews and two literature reviews to investigate the migration from frontend automation with RPA towards backend automation with APIs. This paper explores how the robots need to be designed and prioritized so that they can be easily replaceable in the order that allows an organization to benefit the most from a change in the automation infrastructure. Findings encompass requirements for the design of migration-ready RPA bots, challenges to overcome, and selection criteria for prioritization. We integrate these results into Business Process Management (BPM) from a methodological as well as architectural perspective. The practicability of our findings is backed by a prototypical implementation with which we showcase the migration from frontend to backend automation in an organizational setting from the telecommunications industry.

Keywords: BPM · RPA · APIs · Backend Migration · Design Science

1 Introduction

A vast majority of consumers interacts almost daily with organizations whose legacy infrastructure does not provide open and accessible interfaces. Therefore, many companies move towards Robotic Process Automation (RPA) to tackle the automation of processes using application systems that cannot be integrated otherwise. While RPA is considered useful for rapid automation of these hard-to-integrate components, it is no longer the automation technology of choice once the historically grown systems are given the opportunity of application-internal process automation [2,16]. This requires gradually retiring the robots to replace them with services accessible via Application Programming Interfaces (APIs).

Although RPA has gained momentum in practice, literature does not address how the migration from frontend towards backend automation can look like. Failing to modernize the infrastructure with a more sustainable approach to automation, however, may lead to technical debts for the additional rework at a subsequent time. In response to this research gap, we conduct an exploratory

J. Köpke et al. (Eds.): BPM 2023, LNBIP 491, pp. 149–164, 2023.
https://doi.org/10.1007/978-3-031-43433-4_10

study to analyze how RPA bots can be transitioned towards APIs. First, the design of migration-ready robots is explored so that their latter replacement can happen with ease. Since selecting the wrong candidates can cause automation projects to not deliver the expected results [17], the second goal is to provide strategic foresight in prioritizing robots that allow an organization to benefit the most from a change in the automation infrastructure. Lastly, we implement a prototype to demonstrate the methodological and architectural results regarding the migration-readiness of robots and their prioritized replacement in an industry setting. Thus, the analysis is guided by two research questions (RQs):

1. *How can the development and design of RPA robots be approached when their migration towards APIs is the long-term goal along the automation journey?*
2. *In which order should the migration-ready RPA robots be prioritized so that an organization can get the most out of their transition towards APIs?*

2   Preliminaries

2.1   Business Process Management

BPM provides concepts, methods, and techniques to iteratively improve organizational performance in the form of a lifecycle with circular dependencies [18]. This continuous cycle is structured into the six phases from Fig. 1a [4].

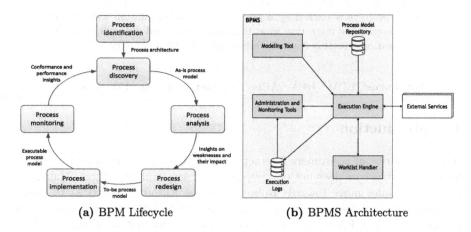

(a) BPM Lifecycle   (b) BPMS Architecture

Fig. 1. BPM Lifecycle and BPMS Architecture [4]

The entry point is the phase of *(1) process identification*. A relevant set of processes is structured into an overall architecture to identify important relationships. Subsequently, *(2) process discovery* intends to document these processes and map the current state on how activities are performed. This typically yields as-is models that are used as input for *(3) process analysis*. The outcome is a collection of existing problems, whose impact is quantified to prioritize which issues should be addressed first. Based on this prioritization, processes are transformed into to-be models during *(4) process redesign*. Due to an abstract design, the to-be models are systematically reworked into executable workflow models along the *(5) process implementation*. Once the redesigned processes are operated, the phase of *(6) process monitoring* is aimed at collecting execution data, observing deviations, and providing the gained knowledge for the next iteration.

Enterprises that practice BPM along this lifecycle may achieve additional benefits if process-aware information systems are used [18]. Such systems are called Business Process Management Systems (BPMSs) and follow the generic architecture from Fig. 1b [4]. This architecture encompasses a *modeling tool* for visualizing process descriptions in terms of their constituent activities. The resultant process models are stored in a *process model repository* so that the *process engine* can pull them for execution. The process engine is responsible for executing the process and distributing the units of work among resources. In case these work items await user interaction, the process engine places them on to worklists for attention by the *worklist handler*. In case the work items require to be completed automatically, *external services* can be consumed that operate outside the realm of the BPMS. This and other execution data is what the process engine keeps *execution logs* of. It allows *administration and monitoring tools* to provide statistics for evaluation with respect to the intended behavior.

2.2 Robotic Process Automation

RPA is a lightweight technology to automate the point of interaction between users and legacy systems [7]. This is realized by mimicking the streams of clicks a user performs in the frontend to provide rapid automation of monotonous activities. It can be used for a broad spectrum of processes, covering a range from single tasks to the automation of entire business processes [1,4]. However, the identification of suitable candidates for automation is crucial and an important step in the beginning of an RPA project [9]. Thus, criteria need to be established to determine which activities are worth automating using RPA [15]. Promising candidates become documented in terms of their behavior in the user interface, developed into RPA bots, and deployed to the respective environment. Subsequently, the robots are tested to ensure they work as expected and monitored to assess the performance in production. As shown in Fig. 2a, these outcomes can be used to enter a new identification cycle for continuous improvement.

In an organizational setting, RPA requires certain architectural necessities to be successfully enacted. As described in Fig. 2b, RPA bots need to orchestrated with the help of a so-called *control system* to manage their execution [11]. This controller holds the repository of the process models that RPA process designers create through interaction with a modeling tool [11]. However, it also distributes the jobs of the process instances among a collection of RPA bots, and exposes both graphical surfaces and programmatic interfaces that an operator can use to start and supervise the ongoing and completed automation jobs [11].

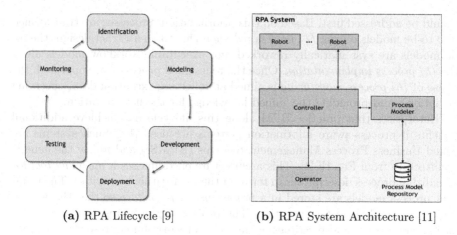

(a) RPA Lifecycle [9]   **(b)** RPA System Architecture [11]

Fig. 2. RPA Lifecycle and RPA System Architecture

3   Methodology

We intend to close the identified research gap with an artefact that is beyond current domain knowledge. To address this objective systematically, the design science research method proposed by HEVNER has established itself as an appropriate approach [8]. For this reason, we structure our research along three related cycles of activities: relevance, rigor, and design.

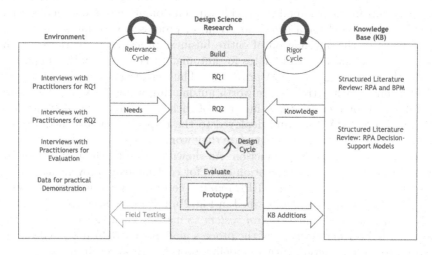

Fig. 3. Adopted Approach to Design Science [8]

Figure 3 presents these cycles together with their circulated domains. As for the first step, we gather input on the basis of interviews with subject-matter experts in an exploratory manner. The collected insights are used to inform the structured literature reviews so that our findings from practice are backed with theory. Subsequently, we conduct the build phase to suggest an initial design for the first and second research question and evaluate its degree of suitability through another round of qualitative interviews. Continuing the evaluation, the improved design is put into a prototype to showcase the migration from frontend towards backend automation in a sample scenario from the telecommunications industry. An evaluation in field conditions remains out of scope for this study.

Table 1. Overview of Interview Partners

Stage	Sector	Organization	Position
1	Insurance	Gothaer Versicherungsbank	Head of RPA
	Financial Services	Commerzbank	Head of Automation
	Engineering	thyssenkrupp Material Services	Project Manager
	Education	Fachhochschule Südwestfalen	RPA Researcher
	Telecommunication	Deutsche Telekom	Vice President IT
			Project Manager
		NetCologne	System Engineer
	IT Consulting	IBM	Platform Architect
			Solution Specialist
		Camunda	Chief Technologist
		Almato	Director Automation
2		WDW Consulting Köln	CEO
		viadee Unternehmensberatung	Managing Consultant

For the exploratory analysis, we approached 11 experts from various industries. A semi-structured technique is selected as the format of the interviews to deviate from the predetermined script whenever we feel the responses require further examination [13]. We stopped the inclusion of additional cases the moment data saturation was reached. The second interview round is conducted with another two process management experts to reassure the rigour of the research. We stick with an unstructured approach to ensure the respondents are not biased by questions set beforehand, rather can pursue the improvement opportunities they feel are important to contribute to an enhanced design [13]. Table 1 provides an overview of sectors, organizations, and positions of all thirteen interviewees we talked to. The choice of respondents is done on the basis of our own judgment and primarily based on hands-on experience with RPA as the leading selection criterion. The interviews took an average of 45 min, were conducted online, and have been analyzed in deductive-inductive manner using the approach by KUCKARTZ for qualitative text analysis [10].

The findings from the first interview round motivate to look up literature at the intersection of RPA and BPM in terms of RQ1. Regarding RQ2, insights reveal the need for consulting decision-support models for the identification of candidates that are suitable for an automation with RPA. For both structured literature reviews, we guide our analysis along the framework proposed by VOM BROCKE ET AL. [3] and query data from five leading scientific databases, namely *Scopus, IEEE Explore, SpringerLink, Web of Science,* and *ACM Digital Library.*

From 216 (166) studies that fit the search string for RQ1[1] (RQ2[2]) initially, 185 (133) are excluded after reading their abstracts. This results in 31 (33) publications that remain for full-text reading, out of which 24 (19) are kept. Without duplicates, this leads to a total number of 19 (11) publications. Through backward referencing, additional 5 (3) articles are identified to be relevant. This gives an overall amount of 24 (14) studies that are left in the final review set.

4   Exploratory Analysis

4.1   Migration-Ready RPA Bots

The analysis of interview data uncovers requirements to the design and challenges to expect during migration. These are summarized in Table 2 and presented next.

Table 2. Design Requirements (DRs) and Migration Challenges (MCs)

ID	Description
DR1	Bots need to automate single tasks on API-granularity in the process
DR2	Bots need to be orchestrated to separate bot- from process-logic
DR3	Bots need to provide automation within the boundary of single systems
DR4	Bots need to be implemented for reuse across processes
DR5	Bots need to be idempotent to realize the same error behavior in the process
DR6	Not only the bot, but also the process logic needs to be designed such that it allows the task to be automated independent of the use of RPA or APIs
MC1	The input and output data between bots and APIs may deviate

Interviewees claim that robots automating tasks within the process are easier to be replaced compared to robots automating the business process itself (cf. DR1). These tasks need to bind modular functionality so that calling a robot to run an operation is conceptually the same as calling an API to realize service behaviour. This requires keeping the process layer apart from the bot logic: instead of confusing a robot's clicks with the essential parts of the process, its activities in a system's user interface should be encapsulated into a service abstraction that is organized along a consolidated workflow (cf. DR2).

It is also recommended to not let a robot interact with multiple systems at once (cf. DR3). Since APIs are built for exposing capabilities from a single core systems, RPA bots should not exceed system boundaries as well. Similarly,

---

[1] ("Business Process Management" OR BPM) AND ("Robotic Process Automation" OR RPA).

[2] ("Robotic Process Automation") AND ("Selection" OR "Identification" OR "Decision Support").

interviewees consider the behaviour on error to be another important factor that needs to be aligned among both approaches to automation (cf. DR5). An API is usually idempotent so that the executions can be repeated without unintended side effects. If RPA bots are integrated into the business process following a different design paradigm, transitioning may become harder as non-idempotent robots may introduce additional process logic that would need to be removed before an API can take over the automation.

Interview data also reveals that the likelihood for a robot to be replaced increases with its reusability across processes (cf. DR4). In case a robot is used only in a single workflow, the migration becomes difficult to justify from a cost perspective. An adaptive design, therefore, allows the transition to pay off faster.

However, the migration capability is said to be not only concerned with the design of the robots themselves, but also with the design of the orchestrating process. Accordingly, the process needs to be implemented so that the business functionality can be handled independently of the choice of the automation technology (cf. DR6). For instance, a process that is designed to always expect an ordered sequence of characters upon a task's completion is going to break in case the API provides the output as an integer value. This requires to refrain from incorporating hard-coded logic that applies for one of the approaches exclusively.

Along the same line of thought, the domain experts expect a major challenge the moment the transition process is triggered. This is due to the fact that the data between frontend and backend is not fully synchronized. Robots may require data as input or output fields that are distinct to the parameters passed as options with the endpoint and vice versa (cf. MC1). This requires the parameters between RPA bots and APIs to be mapped before the migration can begin.

4.2   Migration-Relevant Criteria for Prioritization

Interviewees suggest the prioritization along certain attributes (cf. Table 3). While we discuss their operationalization in Sect. 5.3, this section organizes them into *system characteristics*, *robotic performance*, and *task characteristics*.

In terms of the *system characteristics*, the user interface's stability and the expected end of life are mentioned as factors impacting the candidacy decision. A system with frequent changes to the frontend requires the robot to be reworked regularly. APIs, in contrast, are not sensitive to modifications on the presentation layer and should be the preferred pick for automating tasks that involve applications with many changes to the design. Similarly, a system going out of service in the near future may cause robots to become obsolete. To avoid an existing automation to be unexpectedly hit by the system's shutdown, it is suggested to preferably migrate robots that have to be replaced anyway soon.

In terms of the *robotic performance*, many interviewees posit that the amount of time it takes for a robot to fulfill a service provides yet another indication to derive the urgency for transition. If an RPA bot creates long customer waiting times, the company's outside perception is negatively affected and customers may switch to a competitor. The same holds true for both the rate and quality with which the robot performs the automation. Since APIs usually outperform RPA bots in terms of execution time, automation rate, and reliability, a migration should be initiated whenever low latency levels, a high degree of fully automated cases, or an adequate output quality cannot be guaranteed. Another criterion linked to robotic performance is the number of bot runners, i.e., dedicated machines needed to schedule automation jobs against runtime environments. Whenever the amount of bot runners is detected to be high, practitioners recognize their replacement to hold opportunities for potential cost savings.

The horizontal scale of those machines is determined by the number of times the task to be automated is executed. The higher the transaction amount the more bot runners are needed to perform well under an expanding workload. Thus, APIs are found to be the better fit in situations where the volume with which the task requires execution is extraordinary high.

Table 3. Criteria for the Prioritization of RPA Bots

In Favour of a Migration?	Yes	No
System Characteristics		
Frontend Stability	Changing	Stable
End of Life	Soon	Permanent
Robotic Performance		
Customer Waiting Time	High	Low
Automation Rate	Low	High
Quality of Results	Inadequate	Adequate
Number of Bot Runners	Many	Few
Task Characteristics		
Execution Frequency	Very High	Moderate to High
Number of Systems	One	Multiple
Business Impact	High	Low
Regulatory Compliance	Yes	No
Frequency of Reuse	High	Low

Further *task characteristics* are related to the business impact, the degree to which it can be reused, and the number of systems it interacts with. Particularly, business-critical tasks should be automated with more resilient solutions so that the risk of failure is minimized. Thus, an RPA bot automating a mission-critical activity is more urgent to migrate than a robot automating low-value work. In case this activity is also reusable across various verticals of the organisation, the robot should move up even further in the prioritization hierarchy since a multitude of cases could take advantage of the provided benefits. However, the API-based approach should not be chosen whenever the task encompasses swivel-chair operations that move data of one closed system into another. Since the benefits of RPA take effect all the more when the number of systems to access increases, transitioning a robot that interacts with a single system provides more value compared to a replacement of a multi-system robot.

Finally, practitioners suggest to refrain from automating tasks with RPA in case regulatory constraints exist. If the company has to demonstrate compliance with legal standards relevant to the automated business function, additional security risks may arise due to the use of data that comes from the frontend.

5   Integration of Findings

5.1   Methodological Implications

In order for companies to undertake a gradual modernization of the automation infrastructure, the analysis has shown that interview partners recommend to approach the design of migration-ready RPA bots task-wise and within processes from start to finish. This would allow to replace these fine-granular robots with more robust API operations on the same granularity level, but would require an orchestration layer that handles RPA bots as part of the end-to-end process in the same way as service-oriented APIs. Literature studies this integration from a methodological perspective through a combination of the lifecycle approaches from both RPA and BPM [5,11]. Particularly, FLECHSIG ET AL. extend the BPM lifecyle to consider RPA for processes that have not been worth automating through traditional BPMS [5]. While this approach adopts RPA for process automation in entirety, KÖNIG ET AL. introduce another model that considers RPA activities as individual tasks [11]. They assume the phases of the RPA lifecycle to be implicitly covered by the surrounding BPM framework, and allow the management of RPA processes on a different level of abstraction with the information available to the higher-scope workflow. This way, one layer exists that takes control of the parent process itself, while another layer micromanages RPA on task-oriented granularity along the automation chain.

Since this approach comes close to the requirements interviewees have posed, we choose this model as our base from which we start building. Thus, we follow the idea of using the existing lifecycle model to BPM, but introduce additional layers for RPA phases, DRs, and MCs to become explicitly linked to the phases of the overall framework from Fig. 1a. Furthermore, we need to revise the lifecycle management of RPA bots to make their potential retirement explicit. For this reason, the RPA lifecycle from Fig. 2a is extended so that robots can move into a migration phase after their performance is monitored.

These modifications lead to the four-layered RPA-aware and -migrating BPM framework that is visualized in Fig. 4. It adopts a subset of phases from the BPM lifecycle as the top-level layer and organizes RPA phases, DRs, and MCs underneath. Interviewees suggest to further distinguish DRs and MCs from a business and technical perspective so that technical matters can be explicitly demarcated from aspects that concern the migration's value proposition. Thus, DR4 is found to be not directly contributing to the migratability of a robot itself, rather motivating its replacement from a purely economical standpoint.

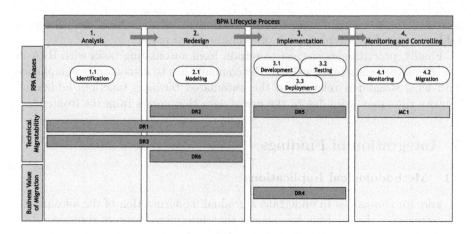

Fig. 4. Framework for the Development of Migration-Ready RPA Bots

The framework's first step encompasses an analysis of an as-is process model through an identification of opportunities for robots to take over tedious tasks that are unable to be automated in the status quo. These tasks should be cut from granularity so that they encapsulate business logic that a service could realize within the boundaries of a single system (cf. DR1 and DR3). Secondly, the process is adapted to convert the manual activities into automated tasks that are part of an outer workflow (cf. DR2). Special care should be taken to ensure this workflow does not include hard-coded process logic that applies for RPA bots or APIs exclusively (cf. DR6). On the RPA layer, the redesign involves capturing the streams of clicks that are necessary to expose modular functionality from the level of the user interface. The resultant RPA model is developed into an idempotent and reusable robot (cf. DR4 and DR5), tested against errors, and deployed to an RPA system in a third step. The overarching workflow, on the other hand, is transformed into an executable process model that is deployed with the help of a BPMS. Since the execution of robots is generally decoupled from the actual orchestration, the BPMS is supposed to hand over control to the RPA system once the process moves on to an RPA-automated activity. The last step is about monitoring process performance. It includes the migration of robots as soon as the need for changing the automation infrastructure is recognized. This requires to prioritize candidates on the basis of their urgency for potential changeover (cf. RQ2), as well as to address the challenge of connecting distinct data models for a substitution with a minimum of effort (cf. MC1).

5.2   Architectural Implications

The development framework implies the orchestration of RPA bots at runtime of a top-level process. In order to ensure the migration-ready RPA bots can be executed as part of this outer workflow, literature argues that BPM and RPA systems need to be integrated so that the units of work can be delegated to a robotic worker whenever needed [6,11]. For this purpose, KÖNIG ET AL. provide an architectural blueprint that uses a BPMS for process orchestration, an RPA system for task automation, and a bridge system that joins them together [11]. Particularly, the bridge system serves as external application that the execution of RPA-automated activities can be delegated to. It forwards the request to the RPA system, waits for the robot to complete the assigned work in the frontend, and returns the result back to the BPMS. While we adopt the idea of having another component that provides access to functionality through the means of a mediator, RQ2 and MC1 yield further requirements that directly translate into architectural necessities. In terms of RQ2, this means that an architectural component is required with which the criteria from Table 3 can be evaluated to rank the robots in the order of their replacement. MC1, on the other hand, motivates the provision of a data mapping so that the transition can happen with the least possible changes to the process logic. These considerations come together to form the migration-enabling architecture depicted in Fig. 5. While we choose *Camunda BPM* as BPMS and *UiPath* as RPA automation software, the architectural idea remains equally feasible with other providers as well.

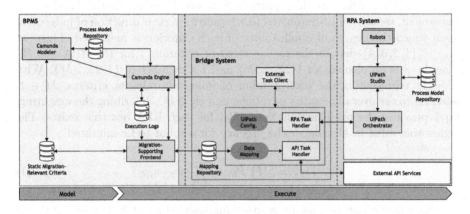

Fig. 5. RPA-Migrating Architecture

Due to the platform's ability to accept extensions, we customize the *Camunda Modeler* so that the static migration-relevant criteria can be evaluated in the time of modeling. This allows the *Migration-Supporting Frontend* to query the criteria from a separate repository and to provide a suggestion for prioritization.

The *External Task Client* is used for realization of the bridge system. As a middleware, it polls the API of the *Camunda Engine* and fetches external tasks to provide them onto worklists. So-called *Task Handlers* pull these work items and process them in asynchronous manner. For the purpose of a migration, two different classes of task handlers are introduced. The RPA task handler is interested in pulling jobs that are meant to be automated by means of robots. It establishes a connection to the RPA system on the basis of access configurations and returns the result upon completion. The API task handler, on the other hand, takes over the automation of external tasks that were formerly intended to be processed by use of RPA. To ensure it can receive and return data in the same way as the robot, the user has to provide mapping templates so that fields from one source can be connected to fields in another source. Thus, the migration from frontend to backend automation becomes a task of publishing the units of work no longer to the worklist the RPA task handler is committed to perform the automation for, rather to the queue the API task handler subscribes to.

5.3 Decision-Support Model for Candidate Selection

Reviewing literature reveals four steps that guide the selection problem between RPA bots and APIs: (1) finding criteria, (2) deriving weights, (3) determining an assessment scale, and (4) collecting data [12,14]. While the migration-relevant information is already defined in Table 3, we extend the architecture from Fig. 5 so that weights can be assigned based on each criterion's relative importance. Thus, we refrain from defining rigid values ourselves, but introduce an additional feature in the *migration-supporting frontend* with which the users can determine a weighting according to their particular needs. In terms of the baseline for assessment, we use a four-numbered likert scale for a clear direction of judgement upon which the degree of applicability of each criterion is measured. Thus, let $v_{jR_i} \in \{1, 2, 3, 4\}$ denote the rating value of criterion j for the RPA bot R_i, whereby j is an element of $\{1, 2, ..., N\}$ and i a member of $\{1, 2, ..., M\}$. With $N = 11$ representing the total number of migration-relevant criteria, $M \in \mathbb{N}$ referring to the overall quantity of robots, and $w_j \in \mathbb{Q}^+$ describing the weighting, Eq. 1 provides the prioritization value P_{R_i} for each RPA bot that exists. The higher this value is, the higher the urgency for a robot to be migrated.

$$\forall i \in \{1, 2, ..., M\} \ P_{R_i} = \sum_{j=1}^{N} (w_j \cdot v_{jR_i}) \tag{1}$$

The collection of data happens either manually through interaction with the modeling tool or automatically by querying the workflow engine. This is, because the criteria can be dynamic or static in nature. Dynamic criteria do not need to be evaluated in advance, rather require evaluation during the execution of the workflow. A robot's automation rate, for instance, should not be assessed when modeling, but needs to be evaluated with each instantiation of the process. This is different with the criteria whose degree of applicability remain static with each new occurrence of the process definition. Citing the examples of the number of systems or the existence of regulatory constraints, it is sufficient to do the evaluation once and neglect the examination at uniform intervals of time.

6   Case Study

We apply the framework for migration-ready RPA (cf. Fig. 4) together with a telecommunications company using the example of a line diagnosis process. As

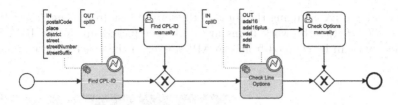

Fig. 6. Redesigned BPMN Diagram of the Line Diagnosis Process

for the first step, we conduct the *process analysis* through an identification of human tasks that are urgent to be robotic-automated. In terms of DR1 and DR3, these activities are decomposed to encapsulate service behavior along the axes of information systems. In this particular case, this includes the tasks of searching the Customer Precise Location-ID (CPL-ID) and assessing various options for the delivery of an internet connection in the designated region. During *process redesign*, both of these user activities are converted into service tasks that are part of an orchestrating workflow (cf. DR1). To further ease migration, we ensure the workflow does not bother with logic dependent on the choice of automation technology (cf. DR6). As a result, Fig. 6 depicts the process after redesign.

With the help of the *Camunda Modeler*, the RPA-configured activities are evaluated in terms of the migration-relevant criteria that is static in nature. For instance, the robot for the search of the CPL-ID is assumed to perform the automation within a system whose user interface is subject to regular change. This assessment can be found in Fig. 7a together with the specification of a topic that an *RPA Task Handler* can use to identify tasks of the same type.

In the next step, the *process implementation* takes place. For this purpose, we develop the RPA bots following DR4 and DR5. While either the search of the CPL-ID and the check of the line options do not manipulate a target resource in the database, the property of idempotency exists by default and does not require further attention during implementation. As for DR4, we make sure that both RPA bots can cover cases that are relevant apart from the process under consideration. For instance, the robot automating the check of the line options is given functionality that allows to not only return the availability of DSL options, but which can also evaluate the status of wireless technologies such as LTE.

Once the RPA bots are deployed to production in reusable and idempotent fashion, the *Camunda Engine* can trigger their execution as soon as the workflow arrives at an RPA-configured task. The *RPA Task Handler* subscribing to the queue called *FindCPLIDRobot*, for instance, reads data from the process context (postal code, place, district, street, number, and suffix), forwards it to the *UiPath Orchestrator*, and provides the result from external processing as CPL-ID.

During *Process Monitoring*, both static and dynamic criteria are evaluated. Assuming the robot that searches the CPL-ID is more urgent to migrate, MC1 is tackled in the belief an API exists that provides the CPL-ID with input fields differing by name. As shown in Fig. 7b, we establish a mapping template so that the data can be transformed as per the expectations of the technical process design. By changing the topic of the service task from *FindCPLIDRobot* to *FindCPLIDApi*, the work items are no longer picked up by the RPA task handler, but can be claimed from the API worker in the exact same manner.[3]

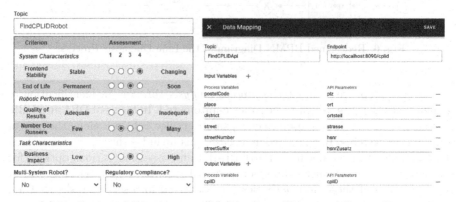

(a) Evaluation of Criteria   (b) Mapping of Input and Output Parameters

Fig. 7. Screenshots of Implementation Artefacts from the Prototype

7 Conclusion

We apply design science research to analyze how the implementation of RPA bots needs to be approached so that they can be easily replaced with APIs. Findings include six requirements to the design of migration-ready robots, eleven criteria with which the robots can be prioritized for transition, and one major challenge that requires particular attention. We organize these findings into a framework that integrates BPM and RPA from a methodological perspective. As the technological enabler, we provide an architectural blueprint that can be used to realize the application of the framework in an organizational setting. Finally, we demonstrate the migration towards APIs in a case study from practice.

However, this research is not without limitations. First, the sample size of the second qualitative evaluation is limited. Second, the prototype is implemented solely for illustration purposes, but cannot be brought into production without change. Third, the built artefacts have not been field-tested in an environment of actual use. Overcoming these limitations requires refinement of the framework for the development of migration-ready RPA bots via multiple experiences in multiple projects. Particularly, additional practitioners should be approached along further design cycles to ensure the results are both generally acceptable and complete. Future work could also include the conduction of field-tests so that the prototype is exposed to situations reflecting its intended use. This could uncover vulnerabilities in our architectural and conceptual considerations that may limit

---

[3] For further illustration, the prototype is showcased in a video on YouTube.

the practical applicability of the artefacts on an organizational and technological level. Nevertheless, the framework, the prototypical implementation, and their sample application introduce preparatory work that has not been explored so far. It provides the skeleton for future studies and presents new ground in closing the gap between frontend and backend automation.

References

1. van der Aalst, W.M.P.: Hybrid intelligence: to automate or not to automate, that is the question. Int. J. Inf. Syst. Proj. Manag. **9**(2), 5–20 (2021)
2. Axmann, B., Harmoko, H., Herm, L.-V., Janiesch, C.: A framework of cost drivers for robotic process automation projects. In: González Enríquez, J., Debois, S., Fettke, P., Plebani, P., van de Weerd, I., Weber, I. (eds.) BPM 2021. LNBIP, vol. 428, pp. 7–22. Springer, Cham (2021). https://doi.org/10.1007/978-3-030-85867-4_2
3. Brocke, J.v., Simons, A., Niehaves, B., Riemer, K., Plattfaut, R., Cleven, A.: Reconstructing the giant: on the importance of rigour in documenting the literature search process. In: ECIS 2009 Proceedings, pp. 2206–2217 (June 2009)
4. Dumas, M., Rosa, M.L., Mendling, J., Reijers, H.A.: Fundamentals of Business Process Management. Springer Berlin Heidelberg (2018). https://doi.org/10.1007/978-3-642-33143-5
5. Flechsig, C., Lohmer, J., Lasch, R.: Realizing the full potential of robotic process automation through a combination with BPM. In: Bierwirth, C., Kirschstein, T., Sackmann, D. (eds.) Logistics Management. LNL, pp. 104–119. Springer, Cham (2019). https://doi.org/10.1007/978-3-030-29821-0_8
6. Flechsig, C., Völker, M., Egger, C., Weske, M.: Towards an Integrated Platform for Business Process Management Systems and Robotic Process Automation, pp. 138–153 (Sep 2022)
7. Herm, L.-V., Janiesch, C., Reijers, H.A., Seubert, F.: From symbolic RPA to intelligent RPA: challenges for developing and operating intelligent software robots. In: Polyvyanyy, A., Wynn, M.T., Van Looy, A., Reichert, M. (eds.) BPM 2021. LNCS, vol. 12875, pp. 289–305. Springer, Cham (2021). https://doi.org/10.1007/978-3-030-85469-0_19
8. Hevner, A.R.: A three cycle view of design science research. Scand. J. Inf. Syst. **19**(2), 87–92 (2007)
9. Jimenez-Ramirez, A., Reijers, H.A., Barba, I., Del Valle, C.: A method to improve the early stages of the robotic process automation lifecycle. In: Giorgini, P., Weber, B. (eds.) CAiSE 2019. LNCS, vol. 11483, pp. 446–461. Springer, Cham (2019). https://doi.org/10.1007/978-3-030-21290-2_28
10. Kuckartz, U.: Qualitative Inhaltsanalyse Methoden, Praxis. Computerunterstützung. Beltz Juventa, WeinheimBasel (2016)
11. König, M., Bein, L., Nikaj, A., Weske, M.: Integrating robotic process automation into business process management. In: Asatiani, A., et al. (eds.) BPM 2020. LNBIP, vol. 393, pp. 132–146. Springer, Cham (2020). https://doi.org/10.1007/978-3-030-58779-6_9
12. Langmann, C., Turi, D.: Robotic Process Automation (RPA) - Digitalisierung und Automatisierung von Prozessen. Springer-Verlag GmbH (Sep 2021)
13. Myers, M.D., Newman, M.: The qualitative interview in IS research: examining the craft. Inform. Organiz. **17**(1), 2–26 (2007)

14. Plattfaut, R., Koch, J.F., Trampler, M., Coners, A.: PEPA: Entwicklung eines scoring-modells zur priorisierung von prozessen für eine automatisierung. HMD Praxis der Wirtschaftsinformatik **57**(6), 1111–1129 (2020)
15. Santos, F., Pereira, R., Vasconcelos, J.B.: Toward robotic process automation implementation: an end-to-end perspective. Bus. Proc. Manag. J. **26**(2), 405–420 (2019)
16. Smeets, M., Erhard, R., Kaußler, T.: Robotic process automation—background and introduction. In: Robotic Process Automation (RPA) in the Financial Sector, pp. 7–35. Springer, Wiesbaden (2021). https://doi.org/10.1007/978-3-658-32974-7_2
17. Wanner, J., Hofmann, A., Fischer, M., Imgrund, F., Janiesch, C., Geyer-Klingeberg, J.: Process selection in rpa projects - towards a quantifiable method of decision making. In: International Conference on Information Systems 2019 (ICIS), pp. 1–17 (2019)
18. Weske, M.: Business Process Management: Concepts, Languages. Architectures. Springer, Berlin New York (2007). https://doi.org/10.1007/978-3-662-59432-2

Accelerating the Support of Conversational Interfaces for RPAs Through APIs

Siyu Huo[1(✉)], Kushal Mukherjee[2], Jayachandu Bandlamudi[2], Vatche Isahagian[3], Vinod Muthusamy[4], and Yara Rizk[3]

[1] IBM Research, Yorktown Heights, NY, USA
siyu.huo@ibm.com
[2] IBM Research, Gurgaon, Haryana, India
{kushmukh,jay_bandlamudi}@in.ibm.com
[3] IBM Research, Cambridge, MA, USA
{vatchei,yara.rizk}@ibm.com
[4] IBM Research, Austin, TX, USA
vmuthus@us.ibm.com

Abstract. In the business automation world, APIs are everywhere. They provide access to enterprise tools such as customer relationship management solutions, or custom automations such as unattended RPA bots that automate repetitive tasks. Unfortunately, they may not be accessible to the business users who need them but are not equipped with the necessary technical skills to leverage them. Most recently, chatbots are becoming the go-to medium to make automation software accessible to business users. Since API specifications aren't written with a chatbot use in mind, additional work is needed to make APIs accessible through a natural language interface. Making this process scalable to many APIs requires an automated data training pipeline for intent recognition models, a crucial component within chatbots to understand natural language utterances from users. More accurate intent recognition models lead to better user experience and satisfaction. Prior work proposed approaches to extracting intents from OpenAPI specifications. However, the resulting models tend to be brittle due to weaknesses in training data. In this work, we propose a data augmentation approach based on paraphrasing using large language models and propose a system to generate sentences to train intent recognition models. Experimental results highlight the effectiveness of our approach. Our system is deployed in a real world setting.

Keywords: Robotic Process Automation · Conversational Assistant · Language Model · Paraphrase Generation · API Specification

1 Introduction

In the age of the fourth industrial revolution, a digital transformation journey is critical to the survival of many companies. Primary sectors adopted automation much sooner than tertiary sectors (e.g., the service industry) [1] despite the availability of millions of Application Programming Interface (API) endpoints. While RPAs helped increase the adoption of automation in enterprises, there are still untapped segments of the market

J. Köpke et al. (Eds.): BPM 2023, LNBIP 491, pp. 165–180, 2023.
https://doi.org/10.1007/978-3-031-43433-4_11

[2]. One obstacle is the lack of accessible automation, given that these companies do not have the technical expertise in the field of software development and automation [2].

One approach to make automation solutions more accessible, including RPA bots and particularly unattended bots [3], is to wrap their API endpoints as chatbots [4] since interactive automation through natural language can be more intuitive to business users [5]. After all, conversational systems were one of the first forms of automation adopted in the service industry [6]; their popularity stems from their attempt to speak the same language as humans.

Typical methods of building chatbots rely on training a shallow intent classifier, which requires expertise in machine learning and natural language processing, and domain experts to provide training data. Another option would be to incorporate large language models (LLMs) that have performed well on many natural language tasks from intent recognition to text generation [7].

LLMs (with millions and even billions of parameters) trained on trillions of tokens from open domain text have performed well on many natural language tasks from intent recognition to text generation [7,8]. However, automation in enterprise settings requires domain-specific knowledge that is not found in any of the data used to train these models. Therefore, fine-tuning these models or providing them with enough contextual information (e.g., when prompting them) is necessary. Unfortunately, for many businesses (especially small ones), fine-tuning such models is prohibitive due to a lack of resources and/or expertise.

As businesses continuously evolve their solutions, new RPA bots (with corresponding APIs) will be created and old ones deprecated. Ensuring that these updated APIs are accessible to employees by carrying out expensive LLM fine-tuning jobs can become excessively costly to companies. A conversational system more amenable to pluggable APIs by training shallow intent classifiers for each API addresses the issue of training cost but still requires AI experts to build these intent classifiers [4].

In this work, we propose an approach to reduce the dependency on AI experts to train the intent classifiers by leveraging LLMs. Our approach relies on paraphrasing sentences extracted from the OpenAPI specification of these APIs. This allows us to leverage the generative strengths of LLMs offline during the development phase, while eliminating the runtime costs of expensive hardware since shallow models don't require GPUs for inferencing. With a human in the loop (subject matter expert as opposed to AI expert) to review the generated sentence, we show that we can achieve comparable performance to LLMs for intent classification while lowering the barrier for non-AI experts.

The outline of our paper is as follows. In Sect. 2, we review related work from the literature. Section 3 presents our proposed approach before Sect. 4 evaluates its performance compared to other approaches in the literature. We discuss how we deployed our solution in a real-world system in Sect. 5 before concluding with final remarks in Sect. 6.

2   Related Work

We briefly survey existing work on generating natural language interfaces to various applications before focusing on natural language generation techniques for the purpose of intent classification.

2.1   Natural Language Interfaces

There are multiple research efforts to generate natural language interfaces (NLIs) automatically from OpenAPI specifications to wrap APIs function in their application. [9] generated a crude chatbot model from Swagger files. They did not rely on sentence generation and focused on engaging with a *power user* to interact with the system and improve over time before actual deployment. [10] adopted an approach that does not include a human in the loop; instead, they augment the training dataset for training NLU intent classification model with unsupervised and distantly-supervised approaches. They use distance metrics in sentence embedding spaces to find similar sentences to those in the API Specification file. Their approach does not take advantage of modern NLG techniques for data augmentation [11] to improve NLU model within NLIs.

2.2   Data Augmentation for NLP

Data augmentation methods look to increase the diversity of training data without explicitly collecting new data; they fall within three main categories [11]. Rule-based methods are naive and cannot guarantee correctness of grammar and semantics. Interpretation-based methods require changing the intent classifier or modification of data pre-processing which is hard to generalize.

Model-based methods take advantage of deep learning neural network architectures such as LSTM [12] and transformers [13] which pushed the boundaries of NLP technology and enabled the development of new high-performance deep learning models such as ELMO [14], BERT [15], GPT-3 [16] and T5 [17]. These pre-trained models can be customized and fine-tuned for the task of text data augmentation with strong effects on performance since they have achieved high accuracy on predicting language and generating diverse paraphrases [18, 19] of the original samples (which is very relevant to our task).

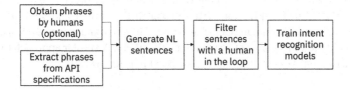

Fig. 1. Intent Recognition Model Training Flow from OpenAPI Specifications

3   Sentence Generation Pipeline

Our goal is to expand the scope of enterprise applications by enabling a conversational interface on top of their available APIs. To achieve that, we need to generate sentences for NLU model training (intent classification). Figure 1 depicts our proposed sentence generation pipeline. We assume that we have access to a few phrases or keywords about the API (either extracted from the OpenAPI specification or provided by the developer of the API). Using those seed phrases, we augment the original sentences with ones generated by deep learning models. Once the sentences are generated, we present the sentences to a human to validate the generated sentences and address model hallucination issues, once the sentences are validated, they become training dataset for training the intent recognition model.

3.1   Key Phrases Extraction Module

Services, exposed as APIs, must provide a specification that adheres to standards (e.g., OpenAPI) to enable consumers of the service to understand and interact with the service with minimal implementation logic. An Example API is shown in Fig. 2. These specifications contain a lot of metadata outlined by keywords. The goal of our key phrase extraction is to utilize this metadata within the API to generate key phrases that can be used to generate the sentences. The API information we are interested in utilizing include: 1) functional and behavioral servers where the resources can be accessed through their endpoints via HTTP protocols, and 2) the descriptions of each endpoint created by the API server developers.

Next, we discuss a few common scenarios of API descriptions and our proposed methods to extract intents from these input descriptions and initiate intent-associated text which is grammatically correct to train the intent classifier.

Scenario 1: Intent from Endpoint Name. In an OpenAPI description, the *path* keyword defines intent-related individual endpoints in the API, and HTTP operations (such as GET, POST) that are supported by these endpoints. Each operation can have some operation objects such as *operationId, summary, description, tags, parameters, requestBody, response* and other metadata.

The *operationId* of each operation is used as the name of the function that handles this request in the generated code, and usually it will contain the action phrase in a certain format which reflects the intent of this function. In order to extract the action phrase, we use regular expressions (RE) to match the format and split word tokens and use context free grammar (CFG) utilizing the verbs and nouns appearing in the tokens to build the full sentence to train intent classifiers. A similar approach to generate sentences based on API descriptions is discussed in prior art [20].

Scenario 2: Intent from Endpoint Description. *Summary* and *Description* keywords contain one or a few sentences which describe the function of the API operation. To utilize the information available in these resources, we first apply semantic parsing (such as abstract meaning representation (AMR) [21] or dependency parsing [22]) on these utterances to extract direct object (dobj) components (e.g., verb-noun phrases).

```
{
  "openapi": "3.0.1",
  "info": {
    "title": "Business Automation Workflow Process REST Interface",
    "version": "201703"
  },
  "paths": {
    "/processes": {
      "get": {
        "tags": [
          "Process"
        ],
        "summary": "Lists process instances.",
        "description": "Lists the process instances that the user may see.",
        "operationId": "listProcessesInstances",
        "metadata": {
          "description": "customer example utterances",
          "content": [
            "Please list process instances for me.",
            "Show me process instances."
          ]
        },
        "parameters": {...}
        "responses": {
          "200": {
            "description": "The request was processed successfully",
            "content": ...
          }
        }
      },
      ...
    }
  }
}
```

Fig. 2. Example OpenAPI specification for an API that lists process instances

Then, we use a back-generator (or CFG) to generate new sentences based on the phrase at the prioritized location in the parsing tree (i.e., close to the root) without any stop words (such as 'is', 'are', 'can' etc.). These new generated sentences are then delivered to the next step in the pipeline.

Scenario 3: Intent from Endpoint Examples. Optional keywords in the API specification give the service developer the ability to define a custom keyword. In particular, the service developer may add a custom ''example utterances'' keyword to the API operation specification and list a few sample sentences that express an intent to invoke this operation. These sentences can be directly used to train the associated intent classifier.

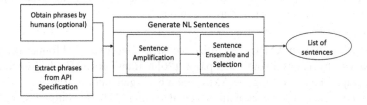

Fig. 3. Sentence Generation Flow

3.2   Sentence Generation Module

Our sentence generation module uses model-based approaches of data augmentation for NLP. Figure 3 shows the main flow of the sentence generation module. Upon receiving a set of sentences from the keyword extraction, we proceed to augment the original sentences with ones generated by deep learning models, then proceed to filter out outliers by taking into consideration both sentence similarity and diversity of generated sentences. Next, we will present the models and approaches we used to generate sentences.

Sentence Amplification. We use the following deep learning methods for paraphrasing the original set of sentences[1]:

Paraphrase with Fine-Tuned T5: The Quora[2] PAWS dataset [23], and The ParaNMT dataset [24] are existing paraphrase datasets for studying paraphrase detection or generation, where an English text is restated using different phrases while maintaining the same meaning. The datasets have the following properties

- **Quora:** The Quora dataset-duplication is composed of 400K pairs of equivalent questions based on real samples from the Quora web site.
- **PAWS:** The PAWS dataset [23] provides around 100K well-formed paraphrase and non-paraphrase pairs with high lexical overlap designed for paraphrase detection. Here, we only use its Wikipedia split.
- **ParaNMT** The ParaNMT dataset [24] provides 50 million English-English sentence paraphrased pairs, which are generated using neural machine translation to translate the Czech side of a large Czech-English parallel corpus.

We tested the T5-base model with default settings and 220M parameters fine-tuned on one of the three paraphrase datasets. Our experiments show that T5-Quora performed better than the other models and is the one we used as the generation model in the pipeline for our deployed system.

Few Shot Learning and Transformation with GPT: GPT3 is an autoregressive language model that performs strongly in a few-shot setting without any gradient updates or fine-tuning. The few-shot learning is done through a prompt, which is a task-specific natural language input, to interact with the model. In our pipeline, we use GPT-Neo, which is a GPT3-like model, and provide paraphrasing requests in the prompt for each input sentence to generate its paraphrased output.

Back-Translation with AMR: We utilize the back-translation [25] framework to generate sentences by: (1) using the AMR parser [26] to transform an English sentence into an AMR graph, then, (2) using the AMR graph back-generator to transform the parsed graphs back to English sentence with diverse decoding [27]. This outputs semantically equivalent but syntactically different sentences. We used Bart-based models with 400M parameters and default settings for each as the parser and back-generator [28].

---

[1] The module can be flexibly expanded with additional models as needed.
[2] https://quoradata.quora.com/First-Quora-Dataset-Release-Question-Pairs.

Sentence Ensemble and Selection. To balance the fidelity and diversity [19] of data augmentation in NLP, the distribution of the generated sentences should neither be too similar nor too different from the original. Otherwise, there is a risk of either overfitting or underfitting the intent classifier by training on non-representative sentences. Thus, we need to consider both fidelity and diversity of the generated sentences used for data augmentation. We collect all the outputs from multiple paraphrasing models as

---

Algorithm 1. Sentence Ensemble and Selection

1: **Input**: Generated sentences S from single model or multi-model(ensemble) in step 2, original sentence s_0, Target size N. Similarity threshold θ, minimum n-gram increasing threshold γ.

2: **Output**: Selected sentences for training intent classifier.

3: Initialization: $G_{selected} \longleftarrow \{\}$

4: Fidelity filtering: $S_{filtered} \longleftarrow \{x | x \in S, Similarity(x, s_0) > \theta\}$

5: **repeat**:

6:     $x_{max} = argmax_{x \in S_{filtered}}(CountNgrams(G_{selected} \cup \{x\}) - CountNgrams(G_{selected}))$

7:     $\Delta ngram = CountNgrams(G_{selected} \cup \{x_{max}\}) - CountNgrams(G_{selected})$

8:     Add x_{max} to $G_{selected}$ **If** $\Delta ngram > \gamma$

9:     Remove x_{max} from $S_{filtered}$

10: **until** $|G_{selected}| =$ N or $|S_{filtered}| = 0$

11: **return** $G_{selected}$

---

candidate sentences. Since these models work in parallel and do not supervise each other, the output paraphrased sentences may be of low quality or have duplicates. For fidelity filtering, we use the MPC-BERT model [29] to generate an embedding of the paraphrased sentences and the original input, and compute the cosine similarity between them[3]. Then, we filter out candidates with similarity values below a threshold θ from the origin. For diversity selection, we prefer paraphrased sentences that have higher unique n-grams and remove the ones with low n-grams [30]. The output sentences are then delivered to the next step in the pipeline section as candidate input training data examples. Algorithm 1 shows the steps to ensemble and select the collected results from these multiple models.

3.3 Interaction with User to Filter Sentences

While the generated sentences have been filtered by our algorithm, we seek user feedback to act as the final arbiter on the quality of the sentences. Thus, in this module, we present our list of generated sentences to the user using a WebUI and ask the users to pick the one they found appropriate. In addition, we provide them with the option of augmenting the generated set of sentences with additional ones. The selected set of user validated sentences are the ones considered in the training data for the NLU intent classifier.

---

[3] Experiments with other similarity metrics such as USE, ROUGE, and BLEU yielded similar results.

3.4   Train the Intent Classifier

Once the training dataset is created, we leverage existing low code chatbot authoring tools like RASA [31] and IBM Watson Assistant (WA) to train the shallow intent classifier. Recent studies [32,33] show that Watson Assistant with its
enhanced intent detection algorithm performs equally well or better than existing intent
classification applications. In this work, we adopt WA as our chatbot authoring tool to
train the intent classifier. Thus, our pipeline evaluation experiments are all conducted
using WA. We believe our results to be representative of other intent classification applications. Because WA is a low-code authoring tool meant to hide the details of training
the models, we do not have access to their hyperparameters to report on in this paper[4].

4   Evaluation

In this section, we evaluate our pipeline on real-world datasets spanning multiple application domains. Our main goal is to establish the effectiveness of our proposed solutions
by (1) evaluating the accuracy of natural language intent classifier trained with sentences generated by our pipeline, (2) understanding the effect of input sentence diversity
and their number on intent classification accuracy, and (3) understanding the effect of
the proposed sentence ensemble and selection method in improving intent classification
accuracy.

4.1   Experimental Setup

We leverage three publicly available intent classification datasets: HWU64 [33], Bank-
ing77 [34], Clinc150 [35], which have been used to benchmark natural language understanding services [32]. The three datasets provide their own training, validation, and
testing splits or cross validation splits which we adopted in our experiments.

HWU64. This dataset provides 64 intents, 21 domains and approximately 25,000
crowdsourced user utterances. The corpus is in the domain of task-oriented conversations between humans and digital assistants [33].

Banking77. This dataset covers around 13,000 customer service queries in the banking
domain labeled with 77 intents. It focuses on fine-grained single-domain intent detection [34].

Clinc150. This dataset consists of around 22,000 examples which includes 150 intent
classes over 10 domains of task-oriented dialog, such as banking, travel, and home. The
dataset also provides out-of-scope examples; however, in this work, we only focus on
in-scope examples [35].

---

[4] http://johncreid.com/wp-content/uploads/2014/12/The-Era-of-Cognitive-Systems-An-
Inside-Look-at-IBM-Watson-and-How-it-Works_.pdf.

Algorithm 2. Sample different groups of representatives for one intent from the training set

1: **Input**: training data D for specific intent.
2: **Output**: $Diverse_n$, $Random_n$, $Narrow_n$ groups, each contain n sentences.
3: Initialization: $Diverse \longleftarrow \{\}$, $Random \longleftarrow \{\}$, $Narrow \longleftarrow \{\}$
4: $Clusters$, $Centers$ = Kmean_cluster (embed(D),n), $|Centers| = n$, $Centers_i \in Clusters$ for $i = 1...n$, $\Sigma_{i \in 1...n}|Clusters_i| = |D|$
5: Smallest cluster: $Clusters_c = argmin_{c \in 1...n}(|Clusters_c|)$
6: $Diverse = \{x_i|x_i = argmin_{x \in D}(\text{distance}(x, Centers_i), \text{for } i = 1...n \}$
7: $Random = \{x_i|x_i = D_{random\_index\_i}, \text{for } i = 1...n \}$
8: **repeat**
9:     $x_{min} = argmin_x((\text{distance}(x, Centers_c))$
    Add x_{min} to $Narrow$
    Remove x_{min} from D for next iteration.
10: **until** $|Narrow| = n$
11: **return** $Diverse_n$, $Random_n$, $Narrow_n$

Since we expect the API developer to provide a few seed phrases or a description of the API. Hence, in our experiments, we sample at most 8 seed phrases for each intent from the datasets as input to our pipeline instead. We conduct our experiments to understand the effect of each component in the pipeline and present important factors that affect the intent classification performance. We note that in this experiment, we bypassed the user feedback phase of the pipeline. Thus, the output sentences are not selected by users before training intent classifier. As indicated previously, we believe that having the user in the loop filtering these sentences will only help to improve the overall accuracy of the classifier.

4.2 Seed Phrase Selection

Recent studies [30] have shown that good user inputs should be diverse enough to cover all aspects of the target intent and thus benefit intent classifier. To choose a diverse set of seed phrases, for each intent in the training data, we apply k-means clustering on the embeddings obtained by using the sentence encoder USE [36] to divide the sentences into n different clusters. For each cluster center, we pick the closest sentence to the center as the representative sentence for this cluster (measured by Euclidean distance of sentence embedding). Thus, n clusters will result in n *diverse representative* sentences (one sentence per cluster) which will be used as user input. In contrast, we pick n sentences around the same center of the smallest cluster (the cluster has the fewest number of examples), and we view these sentences as the *narrow representatives* group. As a baseline comparison, we randomly sample n sentences from the same intent to be used as an example of a user's input, with *narrow* and *diverse* being the worst and best case inputs respectively. Algorithm (2) summarizes our sampling methods.

4.3 Experimental Results

Intent Classification Results

We use the sentences from each of the three groups mentioned above as the seed phrases of our pipeline and configure T5-Quora as the sentence generation model for all three cases. We test the final WA performance where each was trained with the enhanced sentences of the pipeline outputs for one group. Results of the classification accuracy on the test data are shown in Table 1. Note that throughout all experiments, n is the number of input sentences, and the total number of sentences generated using our enhanced pipeline is $5n$.

Table 1 shows that the final intent classification improves significantly when inputs are diverse representative sentences compared to the random case (the former averaged 11.9% better than the latter) and narrow representatives (better by 29.8% on average). Notice that when the input sentence number $n = 1$, the diverse and narrow results are equal (c.f. Table 1) since there is only one cluster. The two methods will pick the same sentence which is the one closest to the cluster center. We discard this case when we compute their average performance difference. We can thus deduce that to improve the overall pipeline performance, it is necessary to collect a diverse set of inputs from the users.

Table 1. Classification accuracy based on input quality using T5-Quora model

HWU64				
Input type	$n = 1$	$n = 2$	$n = 4$	$n = 8$
Diverse	**0.366**	**0.486**	**0.632**	**0.740**
Random	0.174	0.332	0.541	0.658
Narrow	0.366	0.314	0.298	0.361
Bank77				
Input type	$n = 1$	$n = 2$	$n = 4$	$n = 8$
Diverse	**0.420**	**0.565**	**0.718**	**0.782**
Random	0.330	0.384	0.589	0.717
Narrow	0.420	0.346	0.351	0.434
Clinic150				
Input type	$n = 1$	$n = 2$	$n = 4$	$n = 8$
Diverse	**0.490**	**0.679**	**0.795**	**0.862**
Random	0.320	0.546	0.711	0.803
Narrow	0.490	0.420	0.461	0.587

Paraphrasing Model Selection

To choose the appropriate paraphrasing models for various situations, we focus on example utterances that are *diverse* representative of an intent. We select *diverse* as opposed *random* example utterances, because models trained using those sentences performed better (c.f. Table 1). Meanwhile, we use the same model settings for diverse decoding methods, maximum length of output, return sample number, etc. to compare their capabilities.

For each intent, we use the same *diverse* representative sentences as base sentences and configure a pipeline with each candidate sentence generation model. We generate the same number ($5n$) of sentences and train WA models. Table 2 compares the performance of these trained WA models on the testing split for each dataset.

Several papers [35, 37, 38] suggest that data augmentation becomes less beneficial when applied to out-of-domain data, likely because the distribution of augmented data can substantially differ from the original data. Based on the results in Table 2, comparing the performance of three T5 models as sentence generator, we found that T5-Quora performed better than T5-PAWS and T5-ParaNMT. We believe the reason for this is that in task-oriented dialog system, queries in conversations between human and their business consultants (Bank77, Clinic150) or digital assistants (HWU64) are more likely to be expressed as a request or question as in the Quora dataset, rather than statements as in the PAWS (wiki) and ParaNMT datasets. Thus, the T5 model fine-tuned with Quora is closer to the domain of task-oriented dialog and may be able to generate more in-domain sentences and boost the intent classifier's performance.

Table 2. Classification accuracy for different models

HWU64				
Generation model	$n = 1$	$n = 2$	$n = 4$	$n = 8$
T5-Quora	**0.366**	0.486	0.632	**0.740**
T5-PAWS	0.257	0.407	0.605	0.705
T5-ParaNMT	0.237	0.431	0.604	0.716
GPT-Neo	0.341	**0.510**	**0.641**	0.736
AMR	0.306	0.410	0.618	0.711
Bank77				
Model type	$n = 1$	$n = 2$	$n = 4$	$n = 8$
T5-Quora	0.420	**0.565**	**0.718**	**0.782**
T5-PAWS	0.341	0.529	0.687	0.780
T5-ParaNMT	0.357	0.546	0.666	0.778
GPT-Neo	**0.443**	0.553	0.686	0.763
AMR	0.378	0.535	0.699	0.767
Clinic150				
Model type	$n = 1$	$n = 2$	$n = 4$	$n = 8$
T5-Quora	0.490	**0.679**	0.795	**0.862**
T5-PAWS	0.381	0.590	0.791	0.849
T5-ParaNMT	0.401	0.626	0.785	0.844
GPT-Neo	**0.502**	0.678	0.796	0.856
AMR	0.411	0.619	**0.799**	0.845

The results also highlight the extensibility of our pipeline to utilize different types of models: T5, GPT, and AMR for sentence generation.

Table 3. Performance based on pipeline module configuration

HWU64				
Pipeline config	$n = 1$	$n = 2$	$n = 4$	$n = 8$
Base	0.348	0.435	0.588	0.697
Single Model	0.325	0.427	0.621	0.702
Single Model+Selection	0.366	0.486	0.632	0.740
Ensemble+Selection	**0.399**	**0.533**	**0.682**	**0.776**
Parrot	0.351	0.430	0.618	0.711
Banking77				
Pipeline	$n = 1$	$n = 2$	$n = 4$	$n = 8$
Base	0.335	0.516	0.649	0.746
Single Model	0.371	0.532	0.706	0.785
Single Model+Selection	0.420	0.565	0.718	0.782
Ensemble+Selection	**0.482**	**0.599**	**0.731**	**0.787**
Parrot	0.365	0.543	0.677	0.783
Clinc150				
Pipeline config	$n = 1$	$n = 2$	$n = 4$	$n = 8$
Base	0.403	0.605	0.754	0.835
Single Model	0.458	0.619	0.792	0.840
Single Model+Selection	0.490	0.679	0.795	0.862
Ensemble+Selection	**0.535**	**0.698**	**0.821**	**0.866**
Parrot	0.401	0.618	0.763	0.843

Ensemble and Selection

The experimental results above suggest that T5-Quora, GPT-Neo, and AMR perform differently depending on the number of sentences, input quality, and dataset. Taking advantage of model combinations was a motivation for our pipeline design choice. Thus, in this section, we evaluate the scalability of our pipeline to combine results from multiple models. We first examine the effect of using only a single model (T5-Quora) to generate sentence (without enabling sentence selection). Then, we evaluate the effect of using sentence selection on the single model's output (Single Model+Selection), and then an ensemble mechanism that combines the three models' outputs with sentence selection as last component (Ensemble+Selection c.f. Algorithm 1), and this is our final pipeline configurations. In addition, we compare our final pipeline results with a publicly available Parrot[5] paraphrase framework. We use the same *diverse* data as input for

---

[5] Parrot [39] is a paraphrase based phrase augmentation framework specialized in training natural language understanding models. It is based on the HuggingFace T5 library and is in the top 3 downloaded paraphrasing generation models on the website.

our pipeline and Parrot. We tuned Parrot's hyperparameters for each experiment and we report the best scores.

Table 3 highlights results from WA intent classification. The *Base* row lists the results when we use the original n input sentences (representatives) to train the classifier without going through the pipeline nor data augmentation steps (*i.e.* the performance of the prior work which did not generate sentences for data augmentation). The Table shows that by increasingly adding paraphrasing (avg. +2.2%), sentence selection (avg. +3%) and multi-model ensemble modules (avg. +3.1%) into the pipeline, the intent classifier obtains higher accuracy (avg. +8.3%, maximum +14.7%, minimum +3.1% for full integration) than models only trained on the same n input sentences (*Base*). The performance gain is more obvious when the input sentence size n is smaller. Parrot's performance is worse or similar to Single Model+Selection settings. We believe the main reasons for Parrot not performing as well as Ensemble+Selection is: (1) it is designed to only use one paraphrasing model, thus it does not reap the benefits of our ensemble selection which brings higher quality phrases from different models, (2) their diversity ranker and fidelity filter both focus on semantic meaning of sentences and ignore potential lexical diversity of paraphrase [30].

We conduct our experiments using NVIDIA Tesla V100 GPU. On average, the Single Model settings takes at most around 2.0s to generate 10 sentences with 256 tokens, and then the Selection module takes around 0.1s in addition to sorting the 10 sentences by diversity. In our deployed system, the different ensemble models can run in parallel, making it more efficient to generate sentences.

It should be noted that in a few cases, the classifier performance did not improve, even becoming worse, when trained with data from the pipeline (with T5-PAWS, T5-ParaNMT). We checked the outputs and found that in some cases one model may only create limited variations based on its inputs due to the generation model lacking domain knowledge for a specific use case. In such cases, it is important to leverage a human-in-the-loop approach to augment the extracted sentences and improve the generated sentences' diversity.

Error Analysis

The following is an example from the Banking77 dataset for testing the sentence generation pipeline. Given input {*how soon can i get cards?*}, the model generates {*how soon will i get my card? how soon am i supposed to get my card? how soon will i get the ID card?*}. These sentences do add some new n-gram features to the vocabulary and conserve the original meaning, but still lack diverse expressions in the domain and may cause model overfitting on these redundant expressions. The final classifier failed to detect the sentence {*how long will it take to deliver in US?*} from the same intent category (*card_delivery_estimate*) in Banking77. This observation suggests that paraphrasing models have limitations in generating sufficiently diverse examples to cover an intent. We suspect the reason is that the sentence generation models were not specifically fine-tuned with related in-domain data. In the particular example above (from the banking domain), there is a lack of knowledge about card replacement processes which require a delivery system instead of just handing over the card by a neighbor or friend. This can be problematic in some enterprises where getting in-domain data to fine-tune these models may not be feasible. In such cases, it is important to leverage a human-in-the-loop approach to augment the extracted sentences and improve the generated sentences' diversity.

5 Deployed System

The presented approach and a version of our pipeline has been deployed within a real product. The product, IBM's Watson Orchestrate, is based on a multi-agent conversational system [4]. The proposed methodology generates natural language sentences from an OpenAPI specification (see Fig. 4) to train an intent recognition model in IBM's Watson Assistant chatbot authoring tool (see Fig. 5(a)). Once the model is trained, a YAML file defining a conversational agent that includes the generated natural language understanding model in the chatbot and the API endpoint is also automatically generated and activated within the multi-agent system. Finally, a business user can conversationally invoke the desired API endpoint in the chat interface (see Fig. 5(b)).

Fig. 4. Sample OpenAPI specification (description)

(a) Chatbot intent model with generated training examples from API specification

(b) Conversationally invoke the API

Fig. 5. System deployment in production

6   Conclusion

Natural language sentence generation can make chatbot creation (intent recognition models specifically) more accessible to individuals who do not possess the right skills to create them. Our approach proved effective after evaluating it on multiple datasets (accuracy gains were 8.3% on average using diverse representations) and was deployed within a real product. However, the approach does have limitations that we plan on addressing in our future work. This includes taking advantage of domain knowledge when generating sentences and incorporating implicit feedback from users interacting with the chatbot to further augment the training data for intent recognition retraining.

References

1. Laurent, P., Chollet, T., Herzberg, E.: Intelligent automation entering the business world. Deloitte (2015). https://www2.deloitte.com/content/dam/Deloitte/lu/Documents/operations/lu-intelligent-automationbusiness-world.pdf. Accessed 5 Mar 2018
2. da Silva Costa, D.A., São Mamede, H., da Silva, M.M.: Robotic process automation (RPA) adoption: a systematic literature review. Eng. Manage. Prod. Serv. **14**(2), 1–12 (2022)
3. Průcha, P., Skrbek, J.: API as Method for improving robotic process automation. In: Marrella, A., et al. (eds.) Business Process Management: Blockchain, Robotic Process Automation, and Central and Eastern Europe Forum. BPM 2022. Lecture Notes in Business Information Processing, vol. 459, pp. 260–273. Springer, Cham (2022). https://doi.org/10.1007/978-3-031-16168-1_17
4. Rizk, Y., et al.: A conversational digital assistant for intelligent process automation. In: Business Process Management: Blockchain and Robotic Process Automation Forum: BPM 2020 Blockchain and RPA Forum, Seville, Spain, September 13–18, 2020, Proceedings 18, pp. 85–100. Springer (2020). https://doi.org/10.1007/978-3-030-58779-6_6
5. Wang, X., et al: Interactive data analysis with next-step natural language query recommendation. arXiv preprint arXiv:2201.04868 (2022)
6. Behera, R.K., Bala, P.K., Ray, A.: Cognitive chatbot for personalised contextual customer service: behind the scene and beyond the hype. Inf. Syst. Front. 1–21 (2021). https://doi.org/10.1007/s10796-021-10168-y
7. Kenton, J.D., Ming-Wei, C., Toutanova, L.K.: Bert: pre-training of deep bidirectional transformers for language understanding. In: Proceedings of NAACL-HLT, pp. 4171–4186 (2019)
8. Brown, T., et al.: Language models are few-shot learners. Adv. Neural. Inf. Process. Syst. **33**, 1877–1901 (2020)
9. Vaziri, M., Mandel, L., Shinnar, A., Siméon, J., Hirzel, M.: Generating chat bots from web API specifications. In: Proceedings of the 2017 ACM SIGPLAN International Symposium on New Ideas, New Paradigms, and Reflections on Programming and Software, pp. 44–57 (2017)
10. Babkin, P., Chowdhury, M.F.M., Gliozzo, A., Hirzel, M., Shinnar, A.: Bootstrapping chatbots for novel domains. In: Workshop at NIPS on Learning with Limited Labeled Data (2017)
11. Feng, S.Y., et al.: A survey of data augmentation approaches for NLP. arXiv:2105.03075 (2021)
12. Hochreiter, S., Schmidhuber, J.: Long short-term memory. Neural Comput. **9**, 1735–1780 (1997)
13. Vaswani, A., et al.: Attention is all you need. arXiv:1706.03762 (2017)
14. Peters, M.E., et al.: Deep contextualized word representations. In: NAACL (2018)

15. Devlin, J., Chang, M.W., Lee, K., Toutanova, K.: Bert: pre-training of deep bidirectional transformers for language understanding. arXiv:1810.04805 (2019)
16. Brown, T.B., Mann, B., Ryder, N.: Language models are few-shot learners. arXiv:2005.14165 (2020)
17. Raffel, C., et al.: Exploring the limits of transfer learning with a unified text-to-text transformer. arXiv:1910.10683 (2020)
18. Gao, S., Zhang, Y., Ou, Z., Yu, Z.: Paraphrase augmented task-oriented dialog generation. In: ACL (2020)
19. Kumar, A., Bhattamishra, S., Bhandari, M., Talukdar, P.P.: Submodular optimization-based diverse paraphrasing and its effectiveness in data augmentation. In: NAACL (2019)
20. Simsek, U., Fensel, D.A.: Intent generation for goal-oriented dialogue systems based on schema.org annotations. arXiv:1807.01292 (2018)
21. Banarescu, L., et al.: Abstract meaning representation for Sembanking. In: 7th linguistic annotation workshop and interoperability with discourse (2013)
22. Jurafsky, D., Martin., J.H.: Dependency parsing. In Speech and Language Processing (3 ed.). Draft, Chapter 14. https://web.stanford.edu/jurafsky/slp3/14.pdf (2017)
23. Zhang, Y., Baldridge, J., He, L.: Paws: paraphrase adversaries from word scrambling. arXiv:1904.01130 (2019)
24. Wieting, J., Gimpel, K.: ParaNMT-50M: pushing the limits of paraphrastic sentence embeddings with millions of machine translations. In: ACL (2018)
25. Sennrich, R.: Improving neural machine translation models with monolingual data. arXiv:1511.06709 (2016)
26. Bevilacqua, M.: One spring to rule them both: symmetric AMR semantic parsing and generation without a complex pipeline. In: AAAI (2021)
27. Fan, A., Lewis, M., Dauphin, Y.: Hierarchical neural story generation. In: ACL (2018)
28. Lewis, M.: Bart: denoising sequence-to-sequence pre-training for natural language generation, translation, and comprehension. arXiv:1910.13461 (2020)
29. Gu, J.C., Tao, C., Ling, Z., Xu, C., Geng, X., Jiang, D.: MPC-BERT: a pre-trained language model for multi-party conversation understanding. In: ACL/IJCNLP (2021)
30. Yang, Y., et al.: Generative data augmentation for commonsense reasoning. arXiv: Computation and Language (2020)
31. Bocklisch, T., Faulkner, J., Pawlowski, N., Nichol, A.: Rasa: open source language understanding and dialogue management. arXiv preprint arXiv:1712.05181 (2017)
32. Qi, H.: Benchmarking commercial intent detection services with practice-driven evaluations. In: NAACL (2021)
33. Liu, X.: Benchmarking natural language understanding services for building conversational agents. In: IWSDS (2019)
34. Casanueva, I., Temvcinas, T., Gerz, D., Henderson, M., Vulic, I.: Efficient intent detection with dual sentence encoders. arXiv:2003.04807 (2020)
35. Larson, S., et al.: An evaluation dataset for intent classification and out-of-scope prediction. arXiv:1909.02027 (2019)
36. Cer, D.M., et al.: Universal sentence encoder. arXiv:1803.11175 (2018)
37. Campagna, G., Foryciarz, A., Moradshahi, M., Lam, M.S.: Zero-shot transfer learning with synthesized data for multi-domain dialogue state tracking. In: ACL (2020)
38. Zhong, V., Lewis, M., Wang, S.I., Zettlemoyer, L.: Grounded adaptation for zero-shot executable semantic parsing. In: EMNLP (2020)
39. Damodaran, P.: Parrot: Paraphrase generation for NLU. https://github.com/PrithivirajDamodaran/Parrot_Paraphraser (2021)

Exploring Customer Journey Mining and RPA: Prediction of Customers' Next Touchpoint

Jost Wiethölter[1]([⊠]), Jan Salingré[1], Carsten Feldmann[1], Johannes Schwanitz[1], and Jörg Niessing[2]

[1] University of Applied Sciences Münster – Münster School of Business, Münster, Germany
jost.wiethoelter@fh-muenster.de
[2] INSEAD, Fontainebleau, France

Abstract. In-depth analysis of customer journeys to broaden the understanding of customer behaviors and expectations in order to improve the customer experience is considered highly relevant in modern business practices. Recent studies predominantly focus on retrospective analysis of customer data, whereas more forward-directed concepts, namely predictions, are rarely addressed. Additionally, the integration of robotic process automation (RPA) to potentially increase the efficiency of customer journey analysis is not discussed in the current field of research. To fill this research gap, this paper introduces "customer journey mining". Process mining techniques are applied to leverage digital customer data for accurate prediction of customer movements through individual journeys, creating valuable insights for improving the customer experience. Striving for improved efficiency, the potential interplay of RPA and customer journey mining is examined accordingly. The research methodology followed is based on a design science research process. An initially defined customer journey mining artifact is operationalized through an illustrative case study. This operationalization is achieved by analyzing a log file of an online travel agency functioning as an orientation for researchers and practitioners while also evaluating the initially defined framework. The data is used to train seven distinct prediction models to forecast the touchpoint a customer is most likely to visit next. Gradient-boosted trees yield the highest prediction accuracy with 43.1%. The findings further indicate technical suitability for RPA implementation, while financial viability is unlikely.

Keywords: Customer Journey Mining · Customer Journey Mapping · Robotic Process Automation · Process Mining · Prediction

1 Introduction

In recent times, the topic of big data has been attracting an increasing degree of interest. According to the Scopus database, more than 16,000 articles have been published on this subject in the past decade, whereas 13 years ago, less than 100 publications existed on this topic [1]. Alongside this development, the (big) data-based analysis of customer journeys has gained popularity due to its high potential for achieving in-depth

J. Köpke et al. (Eds.): BPM 2023, LNBIP 491, pp. 181–196, 2023.
https://doi.org/10.1007/978-3-031-43433-4_12

customer understanding and enabling an improved customer experience [2]. In this context "customer journey mining" (CJM) was coined, which can be summarized as an interdisciplinary approach that combines concepts from marketing, data analysis, and process management while utilizing process mining techniques to reveal patterns and relationships in the customer journey [3, 4].

The term "customer journey" encompasses the series of interactions and experiences a customer has with a business, from initial awareness through post-purchase evaluation [5]. Customer journeys reflect the customers' perspective, while the term "marketing funnel" describes the company's perspective. Each customer journey consists of different touchpoints, which are defined as the point of contact, where the customer interacts with the companies' products or services [6]. These touchpoints often include multiple systems and service providers offering a seamless, user-centric, and unprecedented range of options, resulting in a so-called omnichannel setup [7]. Customer journeys can be analyzed for various reasons, such as to generate insights into customer satisfaction or factors influencing purchase probabilities [8, 9].

In current business practice, however, customer journey analysis is mostly done retrospectively rather than prospectively, thus only allowing firms to detect anomalies and adapt customer journeys in hindsight [2]. In this context, the ability to proactively and accurately predict each customer's next touchpoint is considered crucial and highly relevant for adapting individual customer journeys, improving the customer experience, and ultimately driving sales [2]. Yet, the absence of formal approaches and adequate tools to analyze and predict individual customer journeys in a target-oriented and automated manner currently poses a hinderance to the practical application [10].

Hence, the main objective of this work is to expand knowledge regarding the possibilities of tool-assisted and data-driven forecasting of customer touchpoints through CJM. To assess the potential of automation in this context, the scope includes exploration of the interplay between robotic process automation (RPA) and CJM for improving efficiency.

To reach the given objective, the paper is structured as follows: Sect. 2 overviews the current field of research by means of a literature review from which the research questions (RQs) are derived. Section 3 then presents the research methodology. In Sects. 4 to 6, the methodology is operationalized through the design of an artifact, which is then demonstrated and evaluated through an illustrative case study showcasing an approach for customer journey mining in combination with RPA. The conclusions of this work and directions for future research are provided in Sect. 7.

2 Research Background

To assess the current state of research and to further sharpen the research agenda, a comprehensive literature review, following the approach of vom Brocke et al. [11] is conducted. The according keyword search targeted towards uncovering customer journey analysis and RPA-related publications is carried out in eight databases and includes titles, abstracts, and keywords within journals and proceedings to increase validity and relevance [12].

As illustrated in Fig. 1, the literature search resulted in 6,994 items, of which 96 papers were selected for further analysis on the basis of criteria currency, relevance, authority, accuracy, and purpose.

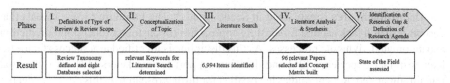

Fig. 1. Literature review process

Current literature exposes a broad range of topics within the field of customer journeys. One core area, which is closely linked to the topic of this paper, is the mapping of customer journeys and their touchpoints. The literature explores different methods and tools to develop detailed and individual views of customer paths. For example, D'Arco et al. [13], and Okazaki and Inoue [14] outline the possibilities of artificial intelligence (AI) and big data, whereas other studies demonstrate and assess process mining techniques to enhance journey mapping [e.g. 6, 15, 16]. Further studies explore the possibilities to analyze specific indicators such as customer satisfaction and loyalty [e.g. 17, 18], churn and conversion rates [e.g. 9, 17] or overall business performance [e.g. 19, 20] through customer journeys. Research has also been conducted with regard to customer touchpoint prediction, reflecting the core part of this paper. Goossens et al. [21] present a process mining-based order-aware recommendation approach (OARA) that allows user paths to be predicted and journeys to be suggested that maximize targeted key performance indicators (KPIs). Further similarly targeted frameworks are the customer journey master approach (CJMA) [22] and the high importance activity prediction (HIAP) [23]. Hassani and Habets [2] present distinct statistical models that leverage log file data to predict future touchpoints.

Overall, the importance of and actual approaches for data-based touchpoint prediction are addressed yet underrepresented throughout literature. Furthermore, the possible integration of RPA in this domain is not being addressed at all, highlighting a clear research gap. Aiming to make an initial step to filling this gap, the following research questions are formulated:

RQ1: How can individual customer touchpoints be predicted using customer journey mining, and what economic benefit can be generated from these predictions?

RQ2: To what extent, if any, can RPA be leveraged to dynamically predict and adapt future touchpoints to improve the individual customer's journey and experience in an automated, efficient manner?

3   Research Methodology

The present work is based on the design science research process (DSRP) model as defined by Peffers et al. [24] and further specified by Hevner et al. [25]. This process model provides a robust and systematic approach for conducting research in the context of information systems and construction-oriented research projects. The DSRP, as outlined by Peffers et al. [24], consists of six phases: problem identification and motivation, objectives of a solution, design and development, demonstration, evaluation and communication. By utilizing the guidelines provided by Hevner et al. [25], sufficient scientific relevance, validation, and effective communication are ensured.

The **problem** area is explained in Sects. 1 and 2: The literature review reveals that current research lacks actual forecasting approaches for customer touchpoint prediction and automated, dynamic processes. As this discipline is identified as a success factor for lean customer experience [26] and decision-support for marketers [13], vast potential remains untapped. Recent studies have called for further research on this topic as a deeper understanding of customer journeys and actual data-driven software-assisted prediction approaches are considered highly relevant for firms and practitioners [7, 27]. Moreover, Romao et al. [28] highlight the need for an increased degree of automation in business processes. The **motivation** of this paper is therefore to further both of these research directions by exploring the interplay between customer journey mining and automation through RPA. The **objectives of the solution** include building on current research; developing a novel, systematic and value-creating approach that serves as orientation for researchers and practitioners and affords them a unified understanding of customer journey mining; and achieving overarching applicability. To **design and develop** the approach by means of an artifact, prior literature is used as groundwork (Sect. 4). The designed approach is **demonstrated** and **evaluated** through an illustrative case study comparing different prediction models and exploring automation opportunities using a log file of a Dutch online travel agency (OTA) as a representative sample data set (Sects. 5 and 6). The automation potential by means of RPA is explored by identifying the individual process steps for modeling and by evaluating them in terms of technical feasibility and financial viability. Lastly, the findings are **communicated** through this paper. Figure 2 summarizes the methodology plan.

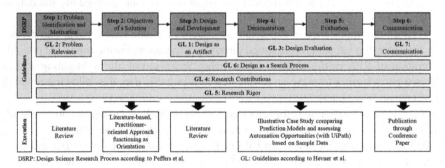

Fig. 2. Methodology plan

4 Design and Development

As a starting point for the design of the artifactual solution, it is to define which dimensions are to consider. These can be derived by the addressed problem area and objectives of the solution in Sect. 3. Overall, the analysis of customer journeys, more specifically, the prediction of future touchpoints, represents the core topic covered in this paper and thus builds the context-giving framework for the desired solution. Since the artifact aims for overarching applicability, the general term "Customer Journey" is selected as dimension one, allowing future applicants to choose the specific focus area they intend to analyze. As the paper aims for tool-assisted and data-based analysis, "big data analytics" is identified as second dimension. To address the identified research gap of the insufficient coverage of RPA in this domain (cf. Sect. 2), "RPA" is defined as third dimension. Consequently, the artifact to be designed is a procedure model for the application of RPA-enhanced customer journey mining.

To achieve this, CRISP-DM (Cross-Industry Standard Process for Data Mining) [29] is incorporated as a procedure model, ensuring the systematic and traceable processing of data while also functioning as groundwork for the artifact (dimension 2). CRISP-DM represents a business-centric standardized model and guideline for data mining activities, which consists of six steps: business understanding, data understanding, data preparation, modeling, evaluation and deployment [29]. Despite CRISP-DM being targeted towards data mining, its application as basis for customer journey mining (process mining-based activity) seems reasonable as data mining serves as foundation for process mining [30] Additionally, previous research projects in this domain have successfully applied CRISP-DM for similar-directed topics [e.g. 2], which underlines its suitability for the given work.

With regards to the possible integration of RPA to enhance efficiency of the process (dimension 3), the requirements for a feasible application are to consider. RPA is an umbrella term for automation tools that imitate human work in the user interface of computer systems [31]. When applied correctly, it can lead to an increase in quality, an increase in process speed, and free capacity of staff [32]. RPA allows to automate individual activities within a process without having to adapt the whole process itself, depicting a clear advantage compared to more traditional automation approaches [33]. As there are certain requirements for a valuable RPA implementation, Langmann and Turi [34] define the following criteria to determine technical suitability for automation (see Table 1):

Table 1. Mandatory criteria for valuable RPA implementation

#	Criteria	Short description
1	Rule-based procedure	RPA-bots are dependent on clear rules for decisions within a process
2	Frequency and repetitiveness	RPA-bots require a high frequency and repetitiveness to create efficiency gains
3	Standardization	RPA-bots require a predetermined, stable process without unpredictable variants
4	Digitalized data	RPA-bots are only capable of reading and using digitalized data

It is evident that a valuable RPA implementation is highly dependent on the individual use case and requires assessment of the criteria presented in Table 1. To cover this prerequisite, these decisive criteria will be included in the design of the artifact.

Based on the three given dimensions, the following artifact is derived (see Fig. 3):

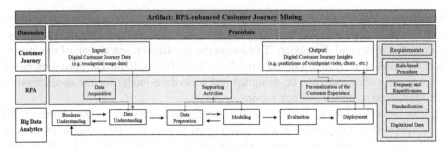

Fig. 3. RPA-enhanced customer journey mining artifact

5 Demonstration

5.1 Overview

The following illustrative case study is based on a log file originating from a Netherlands-based online travel agency (OTA). The data was sourced over a time span of 1.5 years through a user panel in 2015 and 2016, resulting in more than 2 million rows of data. The data collection, capturing digital user journeys, includes information regarding the touchpoints visited as well as the demographic and geospatial characteristics of the panelists. Further details on the data are provided in Sect. 5.3, as the understanding of the data is part of the demonstration of the designed artifact (see Fig. 3).

5.2 Business Understanding

OTAs generate revenue through the bookings made on their websites and the corresponding commissions paid by hotels and airlines for facilitating these transactions [35]. Hence, the **primary aim** of an OTA is to augment the volume of bookings processed through their sales channels. As emphasized in Sect. 1, this objective can be significantly influenced by the analysis and prediction of the customer journey, as it enables retargeting and redirecting customers who, based on the prediction, might leave the customer journey. The improvement of the customer journey through in-depth analysis and prediction can therefore be identified as a secondary objective.

With respect to the **business situation**, OTAs find themselves in a highly competitive market - price advantage, efficiency, system quality, and convenience are critical factors for attracting customers and achieving business success [35]. Notably, the latter three factors are subject to significant influence from the customer journey and its touchpoints, as reflected in the subobjective above.

The **data mining objective,** or in this case, the customer journey mining objective, is therefore to improve comprehension of the models that can predict subsequent customer touchpoints based on different parameters. By achieving this objective, it becomes possible to identify driving factors and subsequently adjust the customer journey and its touchpoints in a data-driven manner.

5.3 Data Understanding

The data collection consists of three files [36]. File A includes data regarding individual touchpoint usage by means of session IDs, user IDs, time stamps, touchpoint types, duration of usage, device used, and purchase completion flag. Overall, this file comprises over 2 million rows of data and covers 20 different touchpoints, ranging from consumer-initiated touchpoints such as Google search to company-initiated touchpoints such as e-mails and prerolls. File B contains user-related data, which covers information about users' gender, age, place of residence, gross income, size of household, education, and occupation. These dimensions are considered particularly relevant as they can later be used in the course of the data analysis to form categories, take subsamples, and highlight interdependencies with regard to the customer journey. File C represents a codebook that is used to uncover the data of Files A and B, which are presented in coded form. Selected demographic and geospatial characteristics of the users are summarized in Fig. 4.

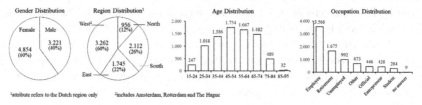

Fig. 4. Overview of selected demographic and geospatial user characteristics

With regard to **data quality**, minor issues can be noted within the customer journey data (File A). The duration attribute of touchpoint usage occasionally manifests as 0 s, implying non-utilization of the touchpoint in such instances. Rows with a usage duration of 0 are consequently eliminated in the preparation phase, not only because they do not offer any added value but also to minimize the likelihood of potential errors in subsequent analyses. Furthermore, the raw data indicates that consecutive clicks on a single point of contact are recorded as individual touchpoints, which deviates from the understanding of touchpoints. Appropriate aggregation for these cases is required.

5.4 Data Preparation

Apart from basic data cleaning, the following preparation steps are applied. First, to link the customer journey data (File A) to the customer-related data (File B), the two data sets are joined based on the *UserID* as the key attribute. Second, all data is decoded based on the codebook (File C). As mentioned above, the touchpoints are not provided at an appropriate granularity. Thus, third, consecutive clicks on the same touchpoint are aggregated based on the time stamps. This step compresses the data from over 2 million rows to approximately 200,000 rows. The data shows that approx. 80% of the customer journeys consist of 10 or less touchpoints visited. Based on this initial preparation step, the customer journey map (Fig. 5) covering the most prominent events can be visualized to enhance understanding of the given use case:

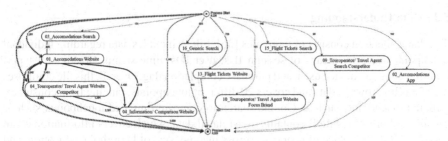

Fig. 5. Visualization of customer journeys

The map indicates which customer journeys are most commonly used (represented by the width of the lines). While such maps do not allow for predictions, they enable an in-depth understanding of customer journeys and aspects such as preferred touchpoints. Here, the touchpoint "01_Accomodations Website" has the highest usage.

Lastly, the arrangement of the available data must be adjusted. The prediction of values, which will be facilitated in the data science platform RapidMiner, is done at column level. Thus, the table layout must be adapted such that each touchpoint has its own column. This is achieved by pivoting the column *touchpoint type* based on the time stamps to keep the correct order. The data set is consequently further compressed to approximately 25,000 rows, representing the data set, File D [36], to be used for prediction.

The predicted variable is the data in column *Touchpoint 3*. This allows to predict future touchpoints in the early stage of a customer journey. As independent variables, the columns containing the demographic and geospatial data as well as the two touchpoints prior to *Touchpoint 3* are taken into account. Variables indicating low correlation with the predicted variable (below 1%), are excluded for modeling (e.g. size of household and size of municipality). Ultimately, 7 independent variables are considered within the prediction models to predict the third touchpoint, as summarized in Table 2.

Table 2. Variables included for modeling

Independent variables				Predicted variable
Gender	Region	Occupation	Touchpoint 1	Touchpoint 3
Age Range	Number of Children		Touchpoint 2	

5.5 Data Modeling and Model Evaluation

The prediction is carried out by employing the seven following distinct models that facilitate multiclass classification. Multiclass classification is deemed necessary due to the large number of potential outcomes (namely, 20 different touchpoints).

Logistic regression is implemented using multinomial logistic regression settings, which allows to predict more than two possible outcomes. **Random forest** and **gradient-boosted trees:** To avoid overfitting, the decision tree depth is set to a maximum of 10. To identify the optimal number of included trees in terms of the highest possible prediction accuracy, the models are trained with a varying number of trees, ranging from 2–150 trees. Furthermore, **deep learning,** representing a multi-layer artificial neural network-based model, is applied. Training results show that after 10 epochs of training, the classification error stops improving. In addition, a **generalized linear model, fast large margin** (support vector machine-based), and the **Naïve Bayes classifier** are facilitated for prediction. The original data set is divided into 60% training and 40% testing data. Table 3 summarizes the modeling results:

Table 3. Modeling results

#	Model	Prediction accuracy	Standard deviation
1	Gradient-Boosted Trees	42.9%	±0.5%
2	Fast Large Margin	41.7%	±1.0%
3	Logistic Regression	40.5%	±0.6%
4	Naïve Bayes Classifier	40.4%	±0.7%
5	Generalized Linear Model	40.3%	±0.5%
6	Random Forest	40.2%	±0.6%
7	Deep Learning	39.2%	±0.6%

All models exhibit a prediction accuracy of approximately 40%. Accuracy is calculated by taking the percentage of correct predictions - in this case for the third touchpoint (cf. Table 2) - over the total number of examples. Correct prediction means the value of the prediction attribute is equal to the value of the label attribute. Gradient-boosted trees promises the highest accuracy with 42.9% and are therefore used for further evaluation. For additional optimization, a three-dimensional optimize parameter grid is applied to identify the model parameters with the highest accuracy for this use case. As shown in Fig. 6, this is achieved by using 30 trees with a maximum depth of 4 and a learning rate of 0.1, which improves the given prediction accuracy by 0.2%.

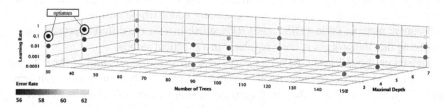

Fig. 6. Prediction model optimization

To put this result into perspective, the prediction accuracy is compared with random guess: randomly predicting one out of 20 possible touchpoints equals a 5% chance of predicting the right one. Thus, the use of the model significantly increases the accuracy of forecasting. Figure 7 depicts an exemplary prediction of the next touchpoint based on the gradient-boosted trees model. The prediction indicates that the given individual customer is most likely to visit the touchpoint "10_Touroperator/ Travel agent Website Competitor" next, within the active customer journey. Based on this insight, various measures could be applied. The customer is likely to visit a competitor's website next, leading to the assumption that the information and offers the customer received at TP 2 were not satisfactory enough to lead to conversion. Thus, a possible measure is an in-depth analysis of this touchpoint with regard to this customer group to identify root causes and potentials for TP improvement. Additionally, the prediction result can be used to trigger individual marketing actions, such as discounts, to redirect the customer.

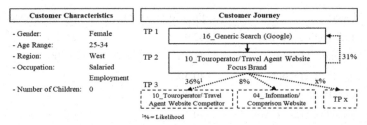

Fig. 7. Exemplary prediction of the next touchpoint

5.6 Deployment

Lastly, the aforementioned model must be deployed. The current procedure for data mining and modeling, as showcased in the previous sections, is mainly based on manual work. As depicted in the artifact (see Fig. 3), the above-applied steps can potentially be covered or supported by RPA if the requirements are met. Till now, only the customer journey mining aspect has been discussed. However, in terms of the overall procedure, the model is to be deployed and the resulting information is to be shared and used accordingly, representing the second core part of this use case. For instance, the customer journey predictions could be leveraged by marketing and UX teams to either target customers and (re-) direct them to preferable touchpoints through adequate measures such as newsletters or ads, or to gradually adapt whole customer journeys. The subsequent matrix (Table 4) illustrates RPA suitability for the different process phases, assuming that the process is done regularly in a practical setting:

Based on these criteria, RPA can cover or support the acquisition of the data, data modeling, and sharing of the generated insights. With regards to the acquisition of new data (1), it is presupposed that the data is conveniently accessible within a centralized repository. At a more detailed level, the introduction of new data files can serve as a triggering mechanism for the initiation of an RPA bot, thereby enabling it to execute the designated task. The operational process entails the bot's ability to navigate to the

Table 4. Assessment of RPA suitability for the given use case

	Rule-based	Frequency & repetitiveness	Standardization	Digitalized data
1 Acquisition of up-to-date data files	✓	✓	✓	✓
2 Data cleaning and preparation	✓	✓	✗	✓
3 Data modeling	✓	✓	✓	✓
4 Genration of prediction results/ reports & analysis	✓	✓	✗	✓
5 Sharing of reports/ insights	✓	✓	✓	✓
6 Retargeting and adaption of touchpoints	o	o	o	o

✓ given   o n/a - depending on business strategy   ✗ not given

specified storage location, retrieve the newly generated file, and subsequently store it in a predetermined destination. Alternatively, the RPA bot can be scheduled to operate at specific time intervals, such as a weekly or monthly basis, irrespective of the availability of new data files. The saved files can then be accessed by the responsible employee for further processing and homogenisation.

The stage encompassing data cleaning and preparation (2), though essential, is regarded as unsuitable for implementation through RPA due to inherent limitations concerning standardization. The inadequacy of standardization stems from the intricate nature of data cleaning, which necessitates detailed examination of progressively complex patterns and unpredictable variants that arise as the dataset expands in size. As a consequence, the application of RPA for this particular task is deemed impractical, as it lacks the flexibility and adaptability required to address the dynamic and evolving nature of data cleaning processes.

With regards to data modeling for this use case (3), RPA can play a pivotal role in facilitating the continuous enhancement and refinement of the initially established model (in this case, gradient-boosted trees). While the initial model holds the potential for predictive purposes, additional value can be derived by incorporating regular updates of new customer journey data. This process involves the continuous integration of fresh touchpoint data, enabling the detection of emerging changes, anomalies, and trends, as well as the evaluation of the efficacy of implemented marketing strategies based on the new prediction outcomes. In order to achieve this, similar to the data acquisition phase, the implementation of an RPA bot can be employed to seamlessly integrate new customer journey data into the underlying dataset of the prediction model as soon as it becomes available. This process is subject to the prerequisite, that clean and prepared data is made available for the bot beforehand.

Similar to the challenges encountered in data cleaning, the level of standardization in utilizing the model for generating prediction results and reports as well as the corresponding analysis (4) is deemed inadequate. The prediction of future touchpoints encompasses a wide array of inquiries and objectives, rendering it a highly individualized process. Consequently, this poses a significant obstacle for the seamless integration of an RPA bot for this process phase.

The subsequent sharing of generated insights/ reports (5) however, could be covered by RPA, as new reports can be used as trigger for the bot to access and upload the files to a predetermined destination or to automatically share it via mail to preset distribution lists. The last process phase, retargeting and customer journey adaptation (6), is considered to be highly individual and depends on the current business strategies, which hinders the implementation of rule-based RPA bots.

Based on the technical feasibility assessment, RPA can be applied to the process phases 1, 3, and 5. As the other phases still require manual work or alternative automation methods, an attended RPA concept, as illustrated in Fig. 8, could be implemented.

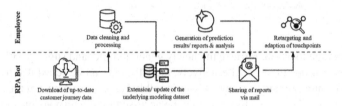

Fig. 8. Attended RPA integration for the given use case

6   Evaluation

In this section, the results of the conducted case study are evaluated against the background of the initially defined objectives of the solution within the DSRP (cf. Sect. 3). Through the demonstration of the developed artifact, it was made possible to observe the extend of its usefulness for efficient customer journey mining in combination with RPA. When it comes to CRISP-DM-based customer journey mining, the artifact offers a reliable procedure to achieve high efficiency and accuracy in terms of predicting future touchpoints along a customer's journey. This part of the artifact thus offers a sound basis and orientation for researchers and practitioners to build on. In order to enhance efficiency, the artifact suggests the application of automation through RPA. Based on the given requirements for a seamless implementation of a reliable RPA bot, it was made possible to accurately identify suitable process phases from a technical point of view. As presented in Sect. 5.6, for the given use case, three process phases can be automated using RPA. However, while technically feasible, it is questionable, whether or not the actual application of RPA is viable from a financial point of view. RPA unfolds its full potential when time-consuming, repetitive tasks are being automated, as this leads to significant time savings for the corresponding employee, who, in return, can spend the time for more complex tasks. However, in the aforementioned process phases, which involve only a limited number of clicks, it is inferred that the potential for significant time savings through the utilization of bots is unlikely. Consequently, the cost-to-benefit ratio for this particular use case is expected to yield a negative outcome. Overall, the artifact achieves the initially defined objectives of the solution by enabling overarching applicability and enabling a unified understanding of customer journey mining in combination with RPA. To further improve validity and practical relevance of the artifact, it is imperative to cover not only technical prerequisites but also requirements assessing financial viability.

7 Conclusion and Future Directions

In this paper, a DSRP-based research project exploring customer journey mining and the potential integration of RPA to predict customer touchpoints was conducted by means of an exhaustive literature review, the development of an artifact and an illustrative case study. Through the development of the artifact and the conduction of the case study based on sample data from a Dutch OTA, RQ1 (How can individual customer touchpoints be predicted based on historical data, and what economic benefit can be generated from these predictions?) can be answered accordingly. By applying CRISP-DM-based data mining techniques as showcased in the previous sections, digital customer journey data can systematically be leveraged to predict future touchpoints. The highest prediction performance is reached through the use of gradient-boosted trees models, achieving a 43.1% accuracy for the given use case. This precision is considered high, as 20 different outcome events are predicted based on 7 parameters. By reducing the number of possible outcomes, which seems reasonable, as not many companies track nor provide 20 distinct touchpoints, the accuracy could be improved even further. The potential economic benefit of forecasting customer touchpoints is rooted in the ability to apply such insights to various domains, such as marketing or UX teams: By having clear visibility into where individual customers are most likely to navigate to next, as opposed to relying solely on qualitative data such as generic persona-based customer journey maps, businesses can proactively respond and implement marketing strategies in a more targeted manner.

However, certain limitations apply. The data used in this case study was collected through a user panel, allowing for precise tracking of each user throughout their customer journeys. In practice, tracking individual users across several touchpoints depicts a key challenge due to data privacy restrictions, insufficient customer tracking opportunities across platforms, or missing data from offline touchpoints. Thus, realizing this potential benefit is highly dependent on the quality of the available data.

Providing an answer to RQ2 (To what extent, if any, can RPA be leveraged to dynamically predict and adapt future touchpoints to improve the individual customer's journey and experience in an automated, efficient manner?), it is evident that RPA can be leveraged to automate several process phases. However, financial viability is highly questionable for the given use case.

The overarching objective of this work, namely to expand knowledge regarding the possibilities of tool-assisted, automated, and data-driven forecasting of customer touchpoints, has been achieved. For future research, it is advisable to extend the developed artifact by a dimension assessing financial viability. Additionally, it seems reasonable to further test the artifact in practical settings in order to generate more insights on its relevance and limitations for the practical application. Furthermore, ways to mitigate the mentioned limitations such as the challenging integration of traceable offline touchpoint data should be explored to further enhance customer journey mining, thereby making it more tangible for future business practices.

Finally, customer journey mining to predict future customer touchpoints represents a vital component within the domain of customer journey analytics.

References

1. Liu, X., Sun, R., Wang, S., Wu, Y.J.: The research landscape of big data: a bibliometric analysis. Libr. Hi Tech **38**(2), 367–384 (2020). https://doi.org/10.1108/LHT-01-2019-0024
2. Hassani, M., Habets, S.: Predicting next touch point in a customer journey: a use case in telecommunication. In: Al-Begain, K., Iacono, M., Campanile, L., Bargiela, A. (eds.) Proceedings of the 35th International ECMS Conference on Modelling and Simulation ECMS 2021, European Council for Modelling and Simulation, Communications of the ECMS, vol. 35, no. 1, pp. 48–54, June 2021
3. Lehnert, F.K.: Customer Journey Mining: Combining User Experience Research Techniques and Data Mining to Capture a Holistic Customer Journey. Technische Universität Eindhoven (2018)
4. Weber, M., Achthoven W., van Slooten, S.: Evidence-based practice in customer experience management: Altuition's customer journey ESPE. In: International Conference on Mass Customization and Personalization – Community of Europe (MCP), pp. 208–215 (2022). https://mcp-ce.org/wp-content/uploads/2022/10/32.pdf
5. Følstad, A., Kvale, K.: Customer journeys: a systematic literature review. J. Serv. Theory Pract. **28**(2), 196–227 (2018). https://doi.org/10.1108/JSTP-11-2014-0261
6. Bernard, G., Andritsos, P.: A process mining based model for customer journey mapping. In: International Conference on Advanced Informations Systems (2017)
7. Barwitz, N., Maas, P.: Understanding the omnichannel customer journey: determinants of interaction choice. J. Interact. Mark. **43**, 116–133 (2018). https://doi.org/10.1016/j.intmar.2018.02.001
8. McColl-Kennedy, J.R., Zaki, M., Lemon, K.N., Urmetzer, F., Neely, A.: Gaining customer experience insights that matter. J. Serv. Res. **22**(1), 8–26 (2019). https://doi.org/10.1177/1094670518812182
9. Lakshminarayan, C., Yin, M.: Topological data analysis in digital marketing. Appl. Stoch. Model. Bus. Ind. **36**(6), 1014–1028 (2020). https://doi.org/10.1002/asmb.2563
10. Halvorsrud, R., Mannhardt, F., Johnsen, E.B., Tapia Tarifa, S.L.: Smart journey mining for improved service quality. In: 2021 IEEE International Conference on Services Computing (SCC), pp. 367–369. IEEE (2021). https://doi.org/10.1109/SCC53864.2021.00051
11. vom Brocke, J., Simons, A., Niehaves, B., Riemer, K., Plattfaut, R., Cleven, A.: Reconstructing the giant: on the importance of Rigour in documenting the literature search process. In: ECIS 2009 Proceedings (2009)
12. Wiethölter, J.: Exploring customer journey mining and RPA: literature review supporting material. Figshare. Online Resour. (2023). https://doi.org/10.6084/m9.figshare.23685123
13. D'Arco, M., Lo Presti, L., Marino, V., Resciniti, R.: Embracing AI and big data in customer journey mapping: from literature review to a theoretical framework. Innovative Market. **15**(4), 102–115 (2019). https://doi.org/10.21511/im.15(4).2019.09
14. Okazaki, K., Inoue, K.: Explainable model fusion for customer journey mapping. Front. Artif. Intell. **5**, 824197 (2022). https://doi.org/10.3389/frai.2022.824197
15. Osman, C.-C., Ghiran, A.-M.: Extracting customer traces from CRMS: from software to process models. Proc. Manufact. **32**, 619–626 (2019). https://doi.org/10.1016/j.promfg.2019.02.261
16. Bernard, G., Andritsos, P.: Discovering customer journeys from evidence: a genetic approach inspired by process mining. In: Cappiello, C., Ruiz, M. (eds.) Information Systems Engineering in Responsible Information Systems. CAiSE 2019. Lecture Notes in Business Information Processing, vol. 350. Springer, Cham (2019). https://doi.org/10.1007/978-3-030-21297-1_4
17. George, M., Wakefield, K.L.: Modeling the consumer journey for membership services. J. Serv. Mark. **32**(2), 113–125 (2018). https://doi.org/10.1108/JSM-03-2017-0071

18. Herhausen, D., Kleinlercher, K., Verhoef, P.C., Emrich, O., Rudolph, T.: Loyalty formation for different customer journey segments. J. Retail. **95**(3), 9–29 (2019). https://doi.org/10.1016/j.jretai.2019.05.001

19. Spiess, J., T'Joens, Y., Dragnea, R., Spencer, P., Philippart, L.: Using big data to improve customer experience and business performance. Bell Labs Tech. J. **18**(4), 3–17 (2014). https://doi.org/10.1002/bltj.21642

20. Rowlson, M.: Uber: process mining to optimize customer experience and business performance. In: Reinkemeyer, L. (eds.) Process Mining in Action. Springer, Cham (2020). https://doi.org/10.1007/978-3-030-40172-6_10

21. Goossens, J., Demewez, T., Hassani, M.: Effective steering of customer journey via order-aware recommendation. In: 2018 IEEE International Conference on Data Mining Workshops (ICDMW), pp. 828–837. IEEE (2018). https://doi.org/10.1109/ICDMW.2018.00123

22. Nguyen Chan, N., et al.: Design and deployment of a customer journey management system: the CJMA approach. In: The 5th International Conference on Future Networks and Distributed Systems, pp. 8–16. ACM (2021). https://doi.org/10.1145/3508072.3508075

23. Wolters, L., Hassani, M.: Predicting activities of interest in the remainder of customer journeys under online settings. In: Montali, M., Senderovich, A., Weidlich, M. (eds.) Process Mining Workshops. ICPM 2022. Lecture Notes in Business Information Processing, vol. 468. Springer, Cham (2023). https://doi.org/10.1007/978-3-031-27815-0_11

24. Peffers, K., et al.: Design science research process: a model for producing and presenting information systems research. Advance online publication (2020). https://doi.org/10.48550/arXiv.2006.02763

25. Hevner, A.R., March, S.T., Park, J., Ram, S.: Design science in information systems research. MIS Q. **28**(1), 75 (2004). https://doi.org/10.2307/25148625

26. Chatzopoulos, C.G., Weber, M.: Digitization and lean customer experience management: success factors and conditions, pitfalls and failures. Int. J. Indust. Eng. Manage. **12**(2), 73–85 (2021). https://doi.org/10.24867/IJIEM-2021-2-278

27. McColl-Kennedy, J.R., et al.: Fresh perspectives on customer experience. J. Serv. Mark. **29**(6/7), 430–435 (2015). https://doi.org/10.1108/JSM-01-2015-0054

28. Romao, M., Costa, J., & Costa, C. J.: Robotic Process Automation: A Case Study in the Banking Industry. In: 2019 14th Iberian Conference on Information Systems and Technologies (CISTI), pp. 1–6. IEEE (2019). https://doi.org/10.23919/CISTI.2019.8760733

29. IBM (Ed.).: IBM SPSS Modeler CRISP-DM Guide (2011). https://www.ibm.com/docs/it/SS3RA7_18.3.0/pdf/ModelerCRISPDM.pdf

30. van der Aalst, W.M.: Process mining in the large: a tutorial. In: Zimányi, E. (eds.) Business Intelligence. eBISS 2013. Lecture Notes in Business Information Processing, vol. 172. Springer, Cham (2014). https://doi.org/10.1007/978-3-319-05461-2_2

31. van der Aalst, W.M.P., Bichler, M., Heinzl, A.: Robotic process automation. Bus. Inf. Syst. Eng. **60**(4), 269–272 (2018). https://doi.org/10.1007/s12599-018-0542-4

32. Smeets, M., Erhard, R.U., Kaußler, T.: Robotic Process Automation (RPA) in the Financial Sector: Technology - Implementation - Success for Decision Makers and Users (1st ed.). Springer eBook Collection. Springer Fachmedien Wiesbaden, Imprint Springer (2021). https://doi.org/10.1007/978-3-658-32974-7

33. Feldmann, C.: Grundlagen zur Automatisierung von Geschäftsprozessen mit Robotic Process Automation. In: Feldmann, C. (eds.) Praxishandbuch Robotic Process Automation (RPA). Springer Gabler, Wiesbaden (2022). https://doi.org/10.1007/978-3-658-38379-4_1

34. Langmann, C., Turi, D.: Robotic Process Automation (RPA) - Digitalisierung und Automatisierung von Prozessen. Springer (2021)

35. Jedin, M.H., Annathurai, K.R.: Exploring travellers booking factors through online booking agency. Int. J. Bus. Inf. Syst. **35**(1), Article 109531, 45 (2020). https://doi.org/10.1504/IJBIS. 2020.109531
36. Wiethölter, J.: Exploring customer journey mining and RPA: case study data. Figshare. Dataset. (2023). https://doi.org/10.6084/m9.figshare.23690811.v1

Educators Forum

Preface

Business Process Management (BPM) has become part of different curricula across primary, secondary, professional, online, and in-company education programs. As educators, we want to pass on our skills and knowledge to new generations. We develop new, often innovative courses with interesting assignments. However, as organizers, we felt that this essential task is often overlooked at conferences. The first edition of the Educators Forum wants to change this, and bring together educators within the BPM community for sharing resources to improve the practice of teaching BPM-related topics by exchanging experiences, bringing in cases, and discussing teaching innovations.

The Educators Forum has received thirteen submissions of which six have been carefully selected based on rigorous reviews for presentation at the forum during the BPM conference.

The paper titled "The Design and Delivery of a Holistic BPM Education Program" reports on the key lessons learned from the authors' experiences conducting BPM education programmes at QUT, and explains how a varied variety of learner-focused technical and managerial factors influenced their BPM education framework.

The paper "Teaching BPM fundamentals: a project-based hands-on course for process-driven systems modeling and development", describes a Business Process Modeling and Automation course built on foundational concepts, showing that the authors have a long experience delivering this course successfully.

In the paper "Digital Case Study: Augmented Process Analysis and Optimization", the authors present a new approach for teaching business process analysis and optimization based on a digital case study. The case study has been elaborated in cooperation with a production company for compression wear, guided by the concept of actions research and the ambition that students can virtually immerse themselves into the company's order-to-cash process.

The paper "Developing industry-ready business analysts: Guiding principles for effective Assessment for Learning (AfL) in a BPM course" discusses a unique assessment approach implemented in the BPM course for Masters' students at the University of South Australia. The authors incorporate Active Learning and Assessment for Learning principles into the course to prepare industry-ready business analysts for the future business environment. The paper discusses the design decisions, implementation, and benefits of three major assignments in the BPM course.

In the paper "Managing BPM Projects and their Implications on Organizations – Experiences from Research and Teaching", the authors report on the use of their tool 'BPM Billboard' in several practitioner workshops and student seminars. They regularly utilize this tool to discuss or showcase BPM initiatives on a high- and holistic level. It is centered around a use case arising from a research project focusing on a real-world application in e-government. In the paper, the authors emphasize that BPM should not be viewed in isolation from the organizational context and that BPM Billboard helps in achieving this goal.

In the study "Bridging the Gap: An Evaluation of Business Process Management Education and Industry Expectations – The case of Poland", the authors have analyzed the gap between the requirements of the labor market and the skills taught in BPM courses at universities in Poland, and propose a standardize BPM course and the introduction of nationwide certification of modeling skills in BPMN.

As can be seen, the Educators Forum addresses a variety of topics on teaching BPM. We want to express our gratitude to all authors who submitted to the forum. All papers were of high quality, and the authors made a hard job for the programme committee to select a programme for presentations. All members have been very helpful during the review process with their detailed and fruitful reviews and comments. We want to thank the organization of BPM 2023, who offered us the opportunity to organize this first Educators Forum. As program committee, we want to dedicate this first edition to Fernanda Gonzalez-Lopez. She was one of the driving forces for organizing this event. Her sudden death has been a terrible shock. It is never going to be the same without her, but we hope that her spirit of enthusiasm, dedication and inspiration will inspire many (new) educators during the Educators Forum, and that the Educators Forum will become the meeting point for sharing and improving our BPM teaching.

September 2023

Katarzyna Gdowska
Fernanda Gonzalez-Lopez (Deceased)
Jorge Munoz-Gama
Koen Smit
Jan Martijn E. M. van der Werf

Organization

Program Chairs

Katarzyna Gdowska — AGH University of Science and Technology, Poland

Fernanda Gonzalez-Lopez (Deceased) — Pontificia Universidad Catòlica de Chile, Chile

Jorge Munoz-Gama — Pontificia Universidad Catòlica de Chile, Chile

Koen Smit — University of Applied Sciences Utrecht, The Netherlands

Jan Martijn E. M. van der Werf — Universiteit Utrecht, The Netherlands

Program Committee

Matthijs Berkhout — University of Applied Sciences Utrecht, The Netherlands

Marlon Dumas — University of Tartu, Estonia

Mahendra Er — Institut Teknologi Sepuluh Nopember, Indonesia

Daniele Grigori — Laboratoire LAMSADE, University Paris-Dauphine, France

Sam Leewis — University of Applied Sciences Utrecht, The Netherlands

John van Meerten — University of Applied Sciences Utrecht, The Netherlands

Jan Mendling — Humboldt-Universität zu Berlin, Germany

Gregor Polancic — University of Maribor, Slovenia

Pascal Ravesteijn — Utrecht University of Applied Sciences, The Netherlands

Manuel Resinas — University of Seville, Spain

Tijs Slaats — University of Copenhagen, Denmark

Mojca Stemberger — University of Ljubljana, Slovenia

The Design and Delivery of a Holistic BPM Education Programme

Rehan Syed⬭, Moe Thandar Wynn⁽⊠⁾ ⬭, and Wasana Bandara⬭

Queensland University of Technology, Brisbane, Australia
m.wynn@qut.edu.au

Abstract. In the dynamic business environment of today, BPM professionals must adapt to cutting-edge technologies, evolving methodologies, and rapidly changing market conditions. Consequently, educational programmes should be sufficiently adaptable to incorporate emerging trends and equip learners with the technical dexterity and process thinking required to effectively lead organisational digital transformation. This adaptability ensures the continued relevance of BPM education and produces graduates that are capable of driving innovation and organisational success.

The goal of this study is to share the key features and report on the key lessons learned from our experiences conducting BPM education programmes for over two decades. We argue for the holistic and flexible nature of BPM education by explaining how a variety of learner-focused technical and managerial factors influence our BPM education framework. We underline the necessity for a well-balanced BPM curriculum that covers both technical topics like process mining and automation and managerial topics like process governance and process strategic alignment. We show how industry and students' needs may be balanced in the BPM curriculum to produce job-ready graduates. This study will contribute to the BPM education community's knowledge base by reporting on the critical dimensions of a programme design that encapsulate our shared experiences in delivering a holistic and flexible BPM curriculum to varied learners using different delivery channels. The details discussed will assist BPM experts in designing and delivering a comprehensive curriculum to tackle BPM programme design and its relevance to today's dynamic business environment.

Keywords: BPM Education · Programme Design · Learner Profile · Programme Theory · Case Contribution

1 Introduction and Background

Business Process Management (BPM) has established itself as an industry best practice to reach an organisation's strategic goals, such as improving customer experience and optimising operational costs through business process efficiencies. BPM's interdisciplinary nature provides a unique opportunity to integrate information technology, data science, and managerial concepts and theories to build robust process solutions and, most importantly, develop an organisation-wide process thinking culture. BPM competencies

J. Köpke et al. (Eds.): BPM 2023, LNBIP 491, pp. 201–210, 2023.
https://doi.org/10.1007/978-3-031-43433-4_13

are highly in demand in Australia, as evidenced by the jobs advertised on the main job search portals. A recent search resulted in 5,917 jobs on Seek and 1,312 jobs on LinkedIn for "Business Process Manager" and 3,065 jobs on Seek and 1,016 jobs on LinkedIn for "Business Process Analyst," indicating a relatively high demand for BPM skills. BPM education equips individuals with the necessary knowledge and skills to analyse, design, implement, and improve business processes, leading to enhanced organisational efficiency and competitiveness. The importance of BPM education is discussed in the literature, highlighting its relevance in today's dynamic business environment. However, vom Brocke [1] stated that due to the rapid development of the field, BPM curricula and training run the risk of rapidly becoming obsolete. To date, there has been limited attention given to BPM education by the BPM community, with only a handful of studies explaining the status, structure, and best practises adopted by BPM education providers [2–7]. Another study [4] suggested that BPM professionals and their employers should build awareness of the existing BPM education options and understand their relevance in meeting industry requirements.

We explain the importance of a comprehensive BPM programme design that includes both technical and managerial competencies in consideration of the industry's multidisciplinary skill demand. We share our insights from delivering BPM education to a diverse range of learners at Queensland University of Technology (QUT), Brisbane, Australia, over the past two decades. QUT was one of the first universities to offer dedicated BPM education courses in the Southern Hemisphere. As of today, only four universities in Australia offer limited BPM units as part of their IT degrees. Globally, UNIR (Spain) is the only university that offers a full BPM programme apart from QUT, whereas a few universities in Europe, the USA, and South America offer BPM education in the form of Postgraduate certificates and CPE.

QUT offers BPM courses to undergraduate Information Technology and Business students, postgraduate Information Technology, Business, and Data Science students, and industry participants. These courses are taught by research-active academics from the Process Science group[1] (formerly the BPM group). The focus of QUT's BPM programme design is on the learners' needs, where individuals seek to acquire a comprehensive understanding of the principles, techniques, and tools required for effective process management through BPM education. We emphasise the need for educational programmes to offer a curriculum that not only addresses the technical aspects of BPM, such as process modelling, improvement, and automation, but also managerial topics, such as governance and strategic alignment. A holistic approach ensures that our graduates have a well-rounded skill set, allowing them to resolve the myriad challenges and opportunities that BPM initiatives present in today's dynamic environment.

This paper unfolds as follows: In Sect. 2, we discussed the key dimensions of QUT's BPM education framework. We emphasised the incorporation of practice-focused, real-world research to enhance the graduates' problem-solving and critical-thinking capabilities. By incorporating research and practice into our programmes, we remain abreast of the most recent BPM developments, industry best practices, and theoretical advances. Section 3 summarises the learners' testimonials and key lessons learned.

---

[1] https://research.qut.edu.au/bpm/

2   Curriculum Design and Delivery

The BPM curriculum at QUT is a carefully curated design that amalgamates several critical elements aimed at heightened levels of pedagogical excellence. The structure is well aligned with programme theory, which advocates the development of a "plausible and sensible model" of how a programme should work by continuously analysing cause & effects relationships between inputs (material, teaching resources, etc.), learners' activities (learning, assessments, etc.), outputs (feedback, results, performance, etc.), and the desired outcomes [8, 9]. This holistic design is the outcome of over two decades of BPM education and continuous reviews and enhancements.

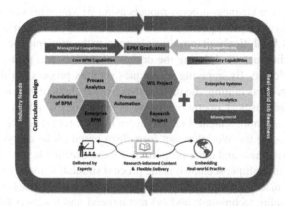

Fig. 1. Overview of the BPM Curriculum Design at Queensland University of Technology

Figure 1 provides an overview of the various dimensions that inform the design and delivery of our curriculum. The course design emphasises the core competencies that a BPM graduate should demonstrate (i.e., process modelling, automation, analytics, and management). These competencies are then demonstrated through real-world practice (an industry project and a research project). Depending on a learner's background knowledge and their desired career path, they can undertake complementary studies in a range of areas. The course is then delivered by experts, both from academia and industry, in a flexible manner using a combination of on-campus, blended, and online modes of delivery.

2.1   Holistic Curriculum Design

The core of the curriculum is made up of four specialised units dedicated to delivering core BPM capabilities. The focused capabilities are determined using feedback from the Industry Expert Group (IEG), alumni, and recruiters at both course and unit levels. These BPM capabilities are then applied through a 'BPM in Practice" project–designed within a work integrated learning (WIL) setup, and a research project. Depending on the learners' profile, they may select from a suite of other unit offerings designed to deliver complementary skills. The number of units depends on the course that they are enrolled in (i.e., a graduate certificate in BPM requires 2 core and 2 elective units, while the master's course requires 4 core, 4 elective, and 2 project units).

The **'Foundations of BPM'** unit provides an in-depth introduction to the management of a business process and takes learners through the fundamental lifecycle phases of a typical business process improvement initiative, from process identification to process monitoring, covering process modelling, analysis, improvement, and automation [10]. On successful completion of this unit, learners will be able to: (i) explain how the business process management lifecycle is applied to guide the management and improvement of organisations' processes; (ii) analyse a business scenario to identify opportunities for business process improvement; (iii) generate and evaluate business process improvement ideas using a variety of quantitative and qualitative analysis techniques; and (iv) create as-is and to-be business process models using BPMN. This unit is the first to be completed and acts as a prerequisite to the other three BPM core units.

The **'Process Analytics'** unit introduces learners to the principles behind various process analysis techniques used during the design, execution, and post-execution stages of the BPM life cycle. By applying techniques such as process verification, process simulation, and process mining, the unit teaches learners how to gain insights into both current and future business operations of an organisation, which in turn can lead to continuous process improvement. Learners learn how to: (i) create executable business process models using the Workflow-net formalism based on an analysis of a domain; (ii) verify the correctness and other properties of these executable models; (iii) evaluate the performance of multiple scenarios using process simulation tools; (iv) analyse existing processes by discovering process models automatically and assess the performance using process mining techniques; and (v) recommend and justify data-driven process optimisation opportunities.

The **'Process Automation'** unit provides a detailed technical and practical exposition of modern business process automation. The learners' learning journey starts with a detailed contextual overview of process automation that also explains the necessity to have executable process instances and how processes can be designed, executed, monitored, analysed, and improved using BPM Systems. Other major themes the unit covers include the theoretical and operational underpinnings of process automation, a detailed discussion on workflow patterns, runtime process flexibility, and exception handling, and the design, implementation, and deployment of process specifications using a state-of-the-art business process automation environment. The unit equips the learners to: (i) apply the theories, strategies, tools, and techniques that inform and support the automation of business processes; (ii) analyse a real-world business scenario to identify opportunities for business process automation; (iii) design and implement automated business processes and supporting components for deployment in process-aware systems environments; (iv) write efficient code for specialised applications that builds services to handle automation tasks; and (v) write user and technical documentation and reports for specialist and non-specialist audiences.

The **'Enterprise BPM'** unit shifts the focus from managing a single process to effectively managing a portfolio of business processes that can be in diverse stages of their respective lifecycles. This unit presents strategic and tactical business and IT management issues with a process-oriented perspective on enterprises and their IT applications.

Based on a comprehensive discussion of key factors surrounding the process-oriented management of an enterprise, BPM capability models are introduced as the baseline tool to use to assess the current status of organisation-wide BPM and to design action roadmaps for BPM capability enhancements across the entire organisation. Learners will be able to: (i) explain why an organisation should adopt a portfolio approach to managing business processes and how to manage BPM portfolios; (ii) select, tailor, and implement BPM capability frameworks to meet the specific needs of diverse organisational contexts; (iii) evaluate the maturity of an organisation's BPM capabilities using capability frameworks; (iv) recommend and justify BPM solutions to a client; and (v) effectively communicate these to an audience that may have little or well-developed BPM skills.

To enable the learners to obtain real-life BPM research and practise skills, they also have to complete a **'BPM in Practice'** project designed within a work-integrated learning (WIL) setup and a **'Research Project'**. With the WIL project, learners have to deliver (pre-agreed) artefacts that will contribute to resolving an issue or realising an opportunity that can be addressed by BPM. The deliverables can be quite diverse; examples include a process architecture for the whole organisation, process improvement recommendations within a set area of the business, a process analytics report for a specific process, or specific algorithms (or algorithmic enhancements) to support process automation efforts. We have dedicated project coordinators whose role entails: (a) managing the learner-educator-industry partner relationships and deliverables; (b) ensuring that the participating case organisation gains value from the engagement; and (c) ensuring that the learner learning outcomes are met. The research projects are also similarly overseen by a dedicated research project coordinator. The research projects fall into two key categories: basic research, where new knowledge is added to the BPM Body of Knowledge by the learner and the supervising educator, or applied research, where the application of existing BPM knowledge into a novel context is studied and evaluated. The research projects often align with QUT's Process Science research group's ongoing portfolio of BPM research activities, where the learner(s) can benefit from a larger ecosystem (e.g., with access to other BPM researchers internally and externally, access to industry partners, and specialised research training and mentoring).

These six units are designed to provide full coverage of the BPM lifecycle. Figure 2 illustrates the alignment of units with the BPM lifecycle. The core units are aimed at developing the foundation for advanced skills, whereas the integrated units provide learners with an opportunity to apply their advanced skills to real-world problems and develop robust solutions.

Finally, learners must also complete at least four approved elective units aimed at equipping them with essential complementary skills. These electives cover Managerial capabilities, such as change management and consultation skills, Enterprise Systems capabilities such as Enterprise Architecture, Enterprise Systems, and Data Analytics capabilities, such as Data analysis for decision making and Foundations of decision science.

	Process Identification	Process Discovery	Process Analysis	Process Re-design	Process Im-plementation	Process Monitoring & Control
Foundations of BPM	Developed	Developed	Developed	Developed		
Process Analytics		Mastered	Mastered			
Process Automation				Mastered	Mastered	
Enterprise BPM	Mastered				Mastered	Mastered
WIL Project	Applied					
Research Project	Applied					

Fig. 2. Mapping with BPM Lifecycle.

2.2 Understanding Industry Needs and Ensuring Real-World Job Readiness

A good education programme must ensure that the contents are aligned with industry needs, delivered by competent facilitators, and provide the learners with an opportunity to apply their knowledge in practical settings. Our learners' real-world job readiness and how well our course can produce graduates who meet industry needs are routinely assessed, with continuous improvements implemented. The curriculum is continually peer-reviewed using multiple feedback mechanisms that include "Industry Expert Group" consultation, learner feedback, and Expert Peer Review (exPrep). We deploy diverse mechanisms such as Focus Groups with practise leaders (both national and international), in-depth interviews and/or focus groups with our alumni and current learners, environmental scans that look at BPM job offerings and industry demand for BPM [11, 12], and comparisons with other BPM course offerings [2]. We also conducted international peer reviews with BPM experts to ensure that our curriculum is current and relevant. We do these as part of our compulsory course review and accreditation cycles, which occur every five (5) years.

2.3 Accommodating Diverse Learner Profiles

When designing educational activities, the use of various elements of a learner's profile can significantly enhance engagement and learning outcomes. Learner profiles encompass an extensive array of characteristics, including learning preferences, multiple intelligences, interests, and strengths. Educators can design learning experiences that accommodate diverse preferences and learning styles by considering these variables. The educators can empower learners to take ownership of their education by considering a variety of learner profiles, such as, interests, career objectives, current skills, and strengths, when designing academic programmes. Learner profiles ultimately aid in the development of successful educational initiatives that promote fruitful learning outcomes and contribute to the overall development of each learner.

QUT's BPM programme has been designed to provide a flexible structure for learners. A BPM professional today is expected to have a strong understanding of both technical and managerial capabilities to lead and contribute to digital transformation initiatives. Learners from technical backgrounds seeking to further their knowledge and skills can select advanced technical units from a wide list of technical units in the Enterprise Systems and Data Analytics clusters. On the other hand, learners seeking to change their career path can develop skills that are more skewed towards managerial topics and have the option to select the required units from the managerial capabilities cluster while maintaining their strong BPM core capabilities. The learning materials are designed using a wide range of techniques, including original case studies, videos, video case studies, industry special seminars, internal research conferences (for research projects), and Work-Integrated Learning projects. The materials can then be consumed by the learner in various manners: face-to-face, hybrid or flip-classroom, hands-on practise, virtual participation, etc.

2.4 Nested Course Structure and Flexible Delivery

The holistic curriculum is complemented by a robust (flexible and nested) course structure to ensure maximum engagement by learners with diverse needs (see Fig. 3). The post COVID-19 era has impacted the design and structure of academic programmes in an unprecedented manner. A specific trend is associated with the high demand for short-term options by industry professionals. Although the short-duration programme offering is not a new concept and there are plenty of examples available, such as MOOCs, edX, and Coursera, their completion rate has been reported to be between 5 and 15% [13]. The Corporate Professional Education pathway allowed learners to complete the required knowledge areas in an accelerated mode (offered in blocks) and receive a certificate of completion upon successfully meeting the required assessment criteria. Once completed, the CPE credits allow learners to join either the Graduate Certificate in BPM or Master of BPM (MBPM) with full recognition of their completed credits.

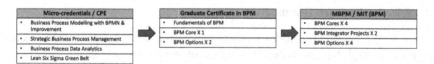

Fig. 3. BPM Course Pathways - A Nested Approach

The Graduate Certificate is offered on a full-time (one semester) or part-time (one year) basis and allows learners to complete an industry-recognised postgraduate qualification. The PGC credentials open the pathway for an advanced entry into the Master of BPM. The Master of BPM is designed to attract learners with prerequisite knowledge of Information Systems or Business Management and is designed to cover the full spectrum of BPM core and complementary skills.

To get the most out of the course, how you deliver the materials is equally important as what learning materials are used in the course. To cater to a variety of learning needs, the courses are offered in a range of modes. On-campus 13-week delivery represents the traditional way of teaching both undergraduate and postgraduate learners, with weekly lectures and tutorials that enable learners to learn and complete hands-on practise. Face-to-face delivery of lectures and tutorials is supplemented by short videos, tool demos, and real-world case studies. Learners will undertake several assessment items, both independently and within a group setting, to demonstrate their grasp of the learning materials. This mode is well suited to full-time learners who are completing a bachelor's or master's degree. Online 10-week + delivery mode is a new mode where learners worldwide can update their knowledge on the fundamentals of BPM. Such a flexible delivery allows interested learners to sample the course content without committing to completing a QUT degree.

To cater for professionals who would like to upgrade their knowledge and capabilities in BPM, we offer the learning materials in block mode as QUTeX continuing professional education offerings. This typically involves a 3- to 4-day intensive short course where groups of 10–15 learners, facilitated by a BPM expert trainer, work through both theoretical and practical components of the learning materials. Additional case studies and homework tasks are set, with assessment items undertaken after the course delivery to help students apply the learning materials at their own pace. This intimate setting with a small group of learners with lived experiences of BPM problems provides rich examples for the learners to share best practises and further enhance their BPM skills.

During and post-COVID-19, such short courses for practitioners are being delivered online only with a mix of synchronous and asynchronous delivery options. For example, a 4-day course is offered as a 2-day course over 2 weeks with the time in between set aside for hands-on practice work. During the day, the course design contains both asynchronous (readings/self-study) and synchronous modes (content delivery/presentations) with discussions and questions via chats and small group meeting rooms supported by technologies such as Zoom.

3 Discussion

3.1 Selective Learner Testimonials

The feedback received from the learners who successfully completed the BPM education programmes at QUT supports the effectiveness of this programme design. For example, the importance of placing **equal emphasis on technical and managerial competencies** has been mentioned as a key feature of the programme by a learner. *"The course really appealed specifically to what I was interested in doing, which was learning about how to link business ideas of a product with an IT solution and how to do that in an efficient manner (MBPM student), it was "…a means to enhance both my technical and management skills in order to drive organisational excellence" (Alumni-1).*

Input design features such as **real-world problem-solving, and industry-informed faculty with practical knowledge of BPM** were well regarded as key factors in meeting the desired outcomes. *"It's terrific getting a connection from theory to real-world practice and being able to ask them questions." (MBPM student). "I enjoyed hearing from industry experts who shared their experiences and insights in a highly relatable manner" (Alumni-2). "The teaching staff and researchers associated with the Master of BPM programme at QUT are outstanding. They not only provided us with theoretical foundations, but also shared practical insights based on their industry and lived experience. Perhaps most notably, the programme consistently challenged us to apply critical thinking and problem-solving skills, which are essential in any technical or leadership role" (Alumni-3).*

The relevance of the learnings to the individual's career progress was highlighted; *"The course gave me insight and knowledge that I can apply in many aspects of my career. I not only learned new skills, techniques, and a better way to think, but I gained an overall confidence which has helped me progress further in my career" (Alumni-4).* The flexible delivery of the units opened more learning opportunities to busy professionals: *"Thank you for running the class in hybrid mode, because of this, I was able to never miss a class despite the several interstate work trips and family challenges" (MBPM student)*; *"Despite not being able to get to Brisbane, I never felt isolated in my learning- there were so many things set up online to help me feel connected/ present and very much part of the class" (MBPM student).*

3.2 Key Lessons Learned

A BPM programme should have a 'holistic curriculum design', which covers skills and competencies required for the various stages of the BPM lifecycle and thus equips the learners with both technical and managerial skills for the continuous management of individual processes and also for process portfolio management. Other aspects pertaining to this principle of holism include career advice through the course coordinator, industry mentors, and formal (e.g., via work-integrated-learning projects) and informal (e.g., via our network and invited guest speakers) mechanisms to assist our graduates in finding BPM job opportunities.

A BPM programme should be **adaptable**—able to cater to emerging industry trends and individual learners career aspects. This requires continuous review and enhancement of the offerings, both in terms of content and delivery. One way we have addressed this at QUT is by trying to maintain a close alignment with our staff's research expertise and teaching content, both of which are driven by contemporary industry demands. We also team-teach and have formal and informal mentoring and coaching in place. We also ensure that students are exposed to various tool sets for modelling, simulation, and process mining.

Industry participation is essential to both the **design** and **delivery** of the curriculum. We accomplished this by establishing a formal "**Industry Advisory Board**," leveraging our staff's extensive industry network, and maintaining close ties with our alumni. These industry representatives play a significant and active role in the course evaluations as well as in the delivery of the learning through guest lectures and projects that require students to 'problem-solve' in the real world.

Flexible delivery is necessary to accommodate the diverse learning preferences of our student populations. Students have requested multiple ways to consume course content, such as (i) on-campus, online, and hybrid modes; (ii) the option to have block-structured offerings (rather than learning over the course of an entire semester); and (iii) units that can yield micro-credentials (so that they can obtain some qualifications as they progress through the course). To adequately resource such diverse learning options, a critical mass of BPM students is required. Therefore, having multiple entry points to the programme(s), as well as a **distinct identity** and **value proposition**, is essential for the programme's long-term success.

References

1. vom Brocke, J.: Where to study business process management? a global perspective based on EDUglopedia. org. Notes. BPTrends, pp. 1–11 (2017)
2. Delavari, H., Bandara, W., Marjanovic, O., Mathiesen, P.: Business process management (BPM) education in Australia: a critical review based on content analysis. In: 21st Australasian Conference on Information Systems, pp. 1–11. AIS Library (2010)
3. Grisold, T., Wurm, B., vom Brocke, J., Kremser, W., Mendling, J., Recker, J.: Managing process dynamics in a digital world: integrating business process management and routine dynamics in IS curricula. Commun. Assoc. Inf. Syst. **51**, 5 (2022)
4. Mathiesen, P., Bandara, W., Marjanovic, O., Delavari, H.: A critical analysis of business process management education and alignment with industry demand: an Australian perspective. Commun. Assoc. Inf. Syst. **33**, 27 (2013)
5. Caporale, T., Citak, M., Lehner, J., Schoknecht, A., Ullrich, M.: Social BPM lab—characterization of a collaborative approach for business process management education. In: 2013 IEEE 15th Conference on Business Informatics, pp. 367–373 (2013)
6. Koch, J., Koch, J., Sträßner, M., Coners, A.: Theory and practice - what, with what and how is business process management taught at German Universities?. In: Di Ciccio, C., Dijkman, R., del Río Ortega, A., Rinderle-Ma, S. (eds.) Business Process Management. BPM 2022. Lecture Notes in Computer Science, vol. 13420. Springer, Cham (2022). https://doi.org/10.1007/978-3-031-16103-2_4
7. Ravesteyn, P., Versendaal, J.: Design and implementation of business process management curriculum: a case in dutch higher education. In: Reynolds, N., Turcsányi-Szabó, M. (eds.) Key Competencies in the Knowledge Society. KCKS 2010. IFIP Advances in Information and Communication Technology, vol. 324. Springer, Heidelberg (2010). https://doi.org/10.1007/978-3-642-15378-5_30
8. Bickman, L.: The Functions of Program Theory. New Directions for Program Evaluation, pp. 5–18. Willy (1987)
9. Pope, A.M., Finney, S.J., Bare, A.K.: The essential role of program theory: fostering theory-driven practice and high-quality outcomes assessment in student affairs. Res. Pract. Assess. **14**, 5–17 (2019)
10. Dumas, M., Rosa, M.L., Mendling, J., Reijers, H.A.: Fundamentals of Business Process Management. Springer (2018)
11. Bandara, W., Rosemann, M., Davies, I., Tan, H.: A structured approach to determining appropriate content for emerging information systems subjects: an example for BPM curricula design. In: 18th Australasian Conference on Information Systems, pp. 1132–1141 (2007)
12. Mathiesen, P., Bandara, W., Delavari, H., Harmon, P., Brennan, K.: A comparative analysis of Business Analysis (BA) and Business Process Management (BPM) capabilities. In: 19th European Conference on Information Systems, pp. 1–13. AIS Electronic Library (2011)
13. eLearning Industry. https://elearningindustry.com/steps-to-boost-your-online-course-completion-rates

Teaching BPM Fundamentals: A Project-Based Hands-On Course for Process-Driven Systems Modeling and Development

Andrea Delgado[✉] and Daniel Calegari

Instituto de Computación, Facultad de Ingeniería, Universidad de la República, Montevideo 11300, Uruguay
{adelgado,dcalegar}@fing.edu.uy

Abstract. Computer Science curricula usually develop software modeling, design, and implementation skills involving standard languages, best practices, and different languages and architectures. However, they do not often involve the specific modeling, design, and implementation of business processes (BPs) using Business Process Management Systems, which require making BPs explicit (e.g., specified in BPMN 2.0) and driving the system development and execution by such models. A specific theoretical, conceptual, and technological background is needed in this context. This paper presents a project-based, hands-on approach to modeling and developing process-driven systems, which we integrated into our Computer Science curricula. We present the experience and highlight lessons learned about essential elements for students learning.

Keywords: Business Process Management (BPM) · Process-Aware Information Systems (PAIS) · project-based teaching · hands-on approach

1 Introduction

Business Process Management (BPM) [1–3] provides support for organizations to focus their daily operation on their Business Process (BPs). Organizations have increasingly adopted BPM in the last two decades. It is promoted by conceptual and technological support, e.g., the OMG standard Business Process Modeling and Notation (BPMN 2.0) [4], and BPM systems (BPMS) [5] providing support for the complete Business Process lifecycle. BPMS as Process-Aware Information Sytems (PAIS) [6] integrate modules supporting process-driven systems, from modeling, configuration, implementation, enactment, and evaluation to improvement. Although BPMS has some differences in implementation, the functionalities offered are mainly conceptually the same [7,8].

A specific theoretical, conceptual, and technological background on BPs is needed to guide this development. Nevertheless, it is only sometimes included in

© The Author(s), under exclusive license to Springer Nature Switzerland AG 2023
J. Köpke et al. (Eds.): BPM 2023, LNBIP 491, pp. 211–221, 2023.
https://doi.org/10.1007/978-3-031-43433-4_14

Computer Science curricula with the essential software systems modeling, design, and implementation. In 2013, we integrated this vision within the undergraduate and postgraduate Computer Science degrees at Universidad de la República in Uruguay. Computer Science is a five-year degree of 450 credits (1 credit equals 15 h of student effort, equivalent to 270 ECTS) organized in semesters. It comprises three basic first years with mandatory courses in mathematics, physics, logic, programming, operating systems, computer architecture, language theory, operations research, numerical methods, computer networks, and databases. The fourth year contains some mandatory courses on software engineering, functional/logic programming, and the final one-year project to graduate. It also offers optional courses covering different areas such as security, advanced computer networks, model-driven engineering, software testing, etc.

This paper presents a project-based, hands-on approach to modeling and developing BPM systems. It consists of a 150 h (6 ECTS credits) optional course of the undergraduate Computer Science degree in the fourth and fifth year, also offered as part of the postgraduate academic Computer Science degree. It is focused on the BPMN 2.0 language for modeling and enacting BPs within an open-source BPMS, following a systematic approach and using best practices for modeling, design, and implementation. We present the course and its evaluation using data from its eight editions. We use the experience to discuss lessons learned, highlighting critical elements for students learning.

The rest of the paper is organized as follows. In Sect. 2, we introduce the course setup, and in Sect. 3, we present our teaching experience. Section 4 discusses results and lessons learned. In Sect. 5, we present related work. Finally, we present the conclusions in Sect. 6.

2   BPM Course Setup

The BPM course is taught in one semester, i.e., all activities are distributed within 15 weeks consisting of classes, laboratories, assignments, and evaluations. We defined some general learning outcomes based on Bloom's taxonomy:

- Acquire basic concepts, techniques, and methodologies to support the BPs lifecycle [**Knowledge**].
- Acquire knowledge on BPs modeling using the standard language BPMN 2.0 and best practices [**Comprehension**].
- Generate experience in BPs modeling and implementation using an open source BPMS platform [**Application**].
- Know and experiment with different BPMS platforms that support the BPs lifecycle with different approaches [**Analysis**].

As depicted in the course schedule of Fig. 1, we divide the course into two main parts with four activity groups: theoretical/practical lessons, hands-on laboratory, hands-on group assignments, and course evaluations. Example materials (Spanish) are here[1]. Weekly lessons and evaluations (each block) are 2-hours

---

[1] Materials: https://www.fing.edu.uy/owncloud/index.php/s/tALdL4dY976v08r.

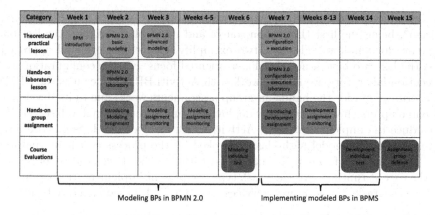

Fig. 1. BPM workshop course typical schedule with lessons categories

activities. The first part (5 weeks plus evaluation) focuses on BPMN 2.0 modeling and best practices. The second part (7 weeks plus evaluation) focuses on implementing the previously modeled process in an open-source BPMS. There is also a general assessment in the form of a group defense of the hands-on work.

Theoretical/practical lessons Since the focus of the course is mostly practical, we introduce key concepts in four weeks of theory/practical lessons for students to get the foundations of BPM knowledge and to perform modeling and development exercises, including:

BPM introduction: we define key concepts of BPM, BP lifecycle (modeling, design, configuration, enactment, evaluation, improvement), BP model, types of BP (collaborative, choreography, orchestration), BP cases, and BPMS.

BPMN 2.0 basic modeling: we introduce the BPMN 2.0 standard and its key elements: activities (tasks and types, sub-processes) and markers (loop, multi-instance, etc.), gateways (AND, XOR, OR), events (start, intermediate, end) and types (message, time, etc.), swimlanes (pools, lanes).

BPMN 2.0 advanced modeling: we introduce and discuss BP modeling best practices: seven process modeling guidelines (7PMG) [9], workflow patterns [10], and process re-design heuristics [11].

BPMN 2.0 configuration & execution: we present key concepts for BPs implementation based on BPMN 2.0 models, which involve designing and developing a software layer to support the BP model execution or high-level elements, depending on the developers-oriented or not focus of the BPMS.

In the practical lessons we delve into most common modeling errors (e.g. granularity of activities), BPMN 2.0 constructs, application of workflow patterns and 7PMG, with a general discussion and key concepts to take away.

Hands-on laboratory The theoretical/practical lessons are supported by two hands-on laboratory lessons using BPMS platforms, allowing students to fix ideas and get involved with the tools firsthand. For the modeling laboratory, we introduce BPMS modeling modules from Activiti[2], Camunda[3], Bonita[4] and Bizagi[5], being the first three open source and multi-platform, and the last one freeware for windows. The first two exemplify a developer-oriented approach, and the last two have a non-developer-oriented focus. For the configuration and execution laboratory, we mainly work with Activiti BPMS since it is the BPMS platform we use for the hands-on group assignments work, due to its developer-oriented approach with Java, easy deployment and execution in a web server. We introduce the implementation in Activiti which involves developing: (a) a Java layer for the BP model with classes invoked by the process engine at runtime, and (b) User task forms to be associated with the user tasks and be presented to the user at runtime. For each task type, we present implementation examples, e.g. service tasks invoking web services with a WSDL-based generated client.

Hands-on group assignments For the group assignment, which is the primary learning element of the course, we follow a project-based hands-on approach, in which groups of 3 or 4 students work together on a project we provide for modeling and implementing a process-driven system in Activiti BPMS. Each year we select two to three real BPs, providing students with a simplified version to work with. The course is online in our Moodle EVA[6] platform, and we use Gitlab[7] as code repository for each group. The group assignment consists of two parts: the first focuses on modeling the BP using BPMN 2.0, and the second on implementing and enacting the BP model in the Activity BPMS platform. We expect students to apply the theoretical/practical knowledge showing the achievement of the defined learning objectives. We include several technological requirements regarding integration aspects: invoking a Web Service from a service task, managing PDF documents with an Electronic Document Management System (EDMS), and use of the Activiti REST API to query process execution.

Course evaluations Regarding the course evaluations, we propose two individual tests, one for each part of the hands-on assignments: one when the BP modeling part ends and the second when ending the development part. Also, the group assignment has a defense at the end of the course. In the individual tests, we ask a few theoretical questions and provide a practical exercise of modeling/development to be solved, similar to the ones we solved in the practical lessons. In the defense, students execute the BP system in Activiti BPMS following a guiding script we provide, with selected data for executing different scenarios, qualifying specific elements of the BP modeling and implementation.

---

[2] https://www.activiti.org/.
[3] https://camunda.com/.
[4] https://www.bonitasoft.com/.
[5] https://www.bizagi.com/.
[6] https://eva.fing.edu.uy/course/view.php?id=423.
[7] https://gitlab.fing.edu.uy/groups/tbpm.

Table 1. Domains and BPs selected for the group assignments

Year	BPs source	Selected BPs
2013–14	University	Teaching position call, Teaching position renewal
2015–16	University	Academic mobility, Agreement management
2017–18	Logistics	Material request, Order dispatch
2019	e-Government	Social benefits allocation, Born alive registration
2021	e-Health	COVID-19 testing, COVID-19 contacts traceability

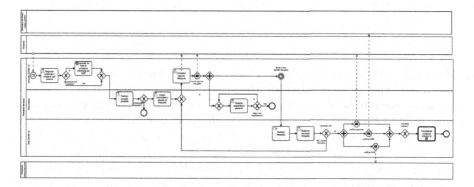

Fig. 2. "COVID-19 testing" e-Health BP excerpt

3   BPM Course Experience

We have taught the course from 2013 to 2019 every year in an in-person format, with hands-on laboratories in computer rooms at the Engineering School; in 2020 and 2022, it was not taught, and in 2021, it was entirely in a virtual format. As mentioned before, we select real BPs for each edition to use as a basis for the project-based group assignment. These BPs come from research projects or collaboration projects, mostly with our country's public organizations, even internal projects from our university that we have carried out. Table 1 presents the BPs used for each year's edition. Figure 2 shows the "COVID-19 testing" BP as an example of their complexity. Other examples in the provided material.

Students face challenges in the assignments associated with our defined learning objectives. Since the course is taught in a Computer Science degree, they have an important background in programming (four prior specific courses and hands-on workshops and laboratories in different programming languages). However, they have only a couple of courses with modeling aspects, mainly with UML and ER models, but none in BPs or BPMN, so it could be expected for them to find more difficult the modeling part. During the course evaluation before 2019, many students expressed that the models were too large and complex. Thus, for the 2019 and 2021 editions, we simplified the control flow and path complexity of the selected BPs, maintaining key elements to learn.

Table 2. Modeling challenging concepts for students

Challenge	Model performed
Control flow collaboration	Message flows as part of the control flow
Control flow orchestration	Missing end events, throw and capture events misused
Participants and roles	Definition of organizational participants/roles as lanes
Granularity of activities	Activities too coarse grained defined
Activity names	Activities with no action names
Type of gateways	Mixed gateway types for diverging and converging flows
Use of gateways	Same gateway for converging and diverging flows
Use of XOR gateways	Missing conditions and corresponding paths

Modeling BPs From the modeling assignment and individual test, we found several modeling elements that are the most challenging for students to understand, as shown in Table 2. Although we provide exercises to illustrate key concepts, including models with errors such as improper use of elements, control flow with inconsistencies, detection of workflow patterns, and modeling best practices, these elements still appear in their solutions. We also found that transitioning from the complete BPMN 2.0 model, e.g., the collaborative model including several pools for different participants, to the single model of each orchestration implemented in the BPMS platform is also challenging.

Implementing BPs From the development assignment and individual test, we found that although most students are well-skilled in Java modeling and development and database definition and management, for information systems development, the process-driven execution in the BPMS platform is a challenging concept to understand. Also, modeling and designing the Java layer to support the process model execution by applying design and architectural patterns to define classes (entities) and associated listeners for execution is not straightforward. Another challenge regarding implementing the BP model is the definition and management of the organizational data model to support the application data not registered in the process engine database. This data model directly relates to the class definition (entities) for the Java layer to manage objects within the process engine execution, as needed by activities, gateways, and other elements within the process control flow.

Course grades The final grade for approving the course comprises two group assignments: BPs modeling and BPs implementation (development) in a BPMS, and two individual tests, one of BPs modeling and another of BPs development. The final grade values obtained by students in all editions are between 7 to 12, with 3 being the minimum grade for approval (60% of the work correctly done).

In Fig. 3, we show the distribution of grades for each assignment in all course editions. The lowest value for the modeling assignment in all editions is 6; for the implementation one, it is 7, and the maximum is 12 for both. In the first four editions (2013 to 2016), the modeling assignment presents lower degrees, both maximum and median, than the development assignment. We did not rank the first assignment in 2017, which included a single delivery. In 2018 there were the worst results in both assignments, mainly related to the complexity and domain of the selected BPs, which were based on real warehouse logistics.

For the last two editions (2019 to 2021), grades are mainly in the same range for both assignments, slightly higher for modeling, being the highest grades for modeling from all editions. We think this results from reducing the complexity of the BPs in these editions, which provided more time to discuss specific modeling aspects reinforcing students' learning. However, this did not impact as expected in the development grades, which remained mainly within the same range in the last four editions, slightly higher in the 2019 edition. In the 2021 edition, the virtual format could have also played a role in this.

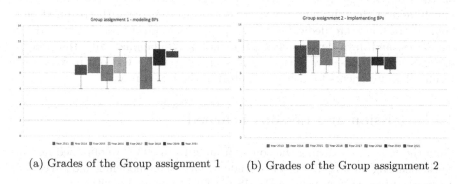

(a) Grades of the Group assignment 1   (b) Grades of the Group assignment 2

Fig. 3. Grades of Group assignments 1 and 2 for all editions

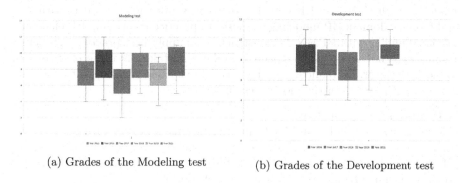

(a) Grades of the Modeling test   (b) Grades of the Development test

Fig. 4. Grades of Modeling and Development test for all editions

Regarding the development assignment and defense with execution in Activiti BPMS, although most groups can execute the scenarios we defined with different datasets and execution paths, some fail for various reasons, including design and technology decisions that groups made independently.

Regarding the modeling and development individual tests, we use them as another control for students within the group assignments work; we introduce the modeling test in the 2015 edition and the development one in the 2016 edition. In Fig. 4, we present the grades for the modeling and development tests from the 2015 edition. It can be seen that although grade ranges are more homogeneous than for the group assignments, minimums are worse in some cases, with values including 2 to 5 for the modeling test and 4 to 6 for the development test. It shows that the learning is not balanced in some groups, and some students are unaware of the modeling and development learning concepts.

Background and hours Around 60% of students work in the software industry, at least 20 h a week and several 40 h or more. They are asked to dedicate 10 h a week to the course, including 2–4 hours for classes and the rest to work on the exercises and assignments. In students' evaluation of the course, they mostly agreed that the hours dedicated are consistent with the required ones. Some extra hours are dedicated to setup the environment for Activiti BPMS, databases, and Java development since they are unfamiliar with the process-driven system perspective. Also, changes in the Activity versions prevent some technical requirements from being fulfilled in the last editions.

4   Discussion and Lessons Learned

From the results of teaching the course from 2013 to 2021, we obtained insights on the challenges Computer Science students face when modeling and developing a process-driven system using a BPMS platform such as Activiti.

Students do not have a background in process-driven systems and modeling languages for processes, but mostly programming in different languages and modeling with UML and ER models. Although we provided students with several familiar domain BP modeling exercises to practice identifying BP elements, become familiar with BPMN 2.0 constructs, and apply best practices, modeling results were not as good as expected, making mistakes as those described in Table 2. However, by reducing the BP's complexity and focusing with them on solving specific parts of the model, results improve.

BP modeling lesson learned 1:
Modeling BPs is challenging for Computer Science students with no process-driven background. Providing a process to model that requires the correct use of key modeling concepts and constructs but with reduced complexity helps them focus on solving a set of specific key challenges.

BP modeling lesson learned 2:
Providing students with the BPMN 2.0 language constructs and reinforcing BP modeling with best practices and workflow patterns does not directly impact the expected results for modeling and development. Students still make basic mistakes which we believe they can improve with practice.

Most students are skilled in Java modeling and development, database definition, and management. Moreover, they have experience in collaborative software development projects for traditional information systems. However, the process-driven execution in the BPMS platform is a challenging concept. Also, the design of the Java layer to support the process model execution and the definition and management of the organizational data model are not straightforward.

BP development lesson learned 1:

Although most students could be well-skilled in programming, they are unfamiliar with process-driven systems, and understanding the process engine and the inversion of control is challenging. Project-based hands-on assignments provide them with a natural and valuable experience developing these systems.

BP development lesson learned 2:

Guides on defining organizational data models to support application data managed by the BP model apart from the process engine database are needed. Understanding how to connect organizational data with the Java layer within the process model execution by the BPMS platform is not straightforward.

The experience yields information that allows some preliminary conclusions to be drawn. However, a more in-depth validation is necessary to transform them into recommendations. For this, we need to reinforce the evaluation of some aspects. In terms of modeling, we could compare the modeling results of groups that use good practices with groups that do not, as well as address domains with different levels of complexity. We could also get an actual client to evaluate the understanding of the models made by the client beyond the modeling errors that we, as teachers, identify. Regarding development, we could determine the use of other platforms that have evolved better, such as Camunda or Flowable.

5   Related Work

There is significant work on process modeling concerning best practices [9], workflow patterns [10], and process re-design heuristics [11], traditionally assessed in theoretical and practical contexts. Most works focus on teaching and learning such modeling concepts. In [12], the authors present an e-learning approach for process modeling education based on requirements derived from related work on modeling and e-learning. In [13], the authors focus on identifying difficulties of teaching/learning state-oriented modeling based on reflections from teachers and learners and a small-scale survey. In [14], the authors describe an introductory BPM course lectured in a college of business administration focused on process modeling and simulation. In [15], the authors present an undergraduate and postgraduate course in which students are asked to conceptualize, analyze, and articulate real-life process scenarios for PAIS design. There are also works about teaching a more general BPM perspective, not only focused on process modeling. In [16], the authors present a disruptive strategy based on role-play using Second Life to introduce students to the properties of ERP systems and simultaneously introduce tools for virtual team collaboration. In [17], the authors report on an analysis of the current BPM offerings of Australian universities. In [18], the authors describe their experience in teaching BPM as a Massive Open Online Course (MOOC) with a comprehensive coverage of the BPM lifecycle.

6   Conclusions

We have presented an undergraduate and postgraduate project-based hands-on course for process-driven systems modeling and development we have taught in a Computer Science degree including: course setup, experience, critical results for the 2013 to 2021 editions concerning the learning objectives, and lessons learned.

Among the challenges for students learning, we have identified that modeling BPs in BPMN 2.0 with best practices is the most challenging task of the course for them. Also, BPs implementation in a BPMS platform poses a critical challenge, mostly how the process engine works, the inversion of control from traditional systems (i.e., work list), and modeling and designing the Java layer and data model to support the process. We have taken a few actions to ease students learning for the challenges identified, from which some have already impacted in a positive way. Further evaluation is needed to deepen the results.

References

1. Weske, M.: BPM - Concepts, Languages, Architectures, 3rd Ed. Springer (2019). https://doi.org/10.1007/978-3-662-59432-2
2. Dumas, M., Rosa, M.L., Mendling, J., Reijers, H.A.: Fundamentals of BPM, 2nd Ed. Springer (2018). https://doi.org/10.1007/978-3-662-56509-4
3. van der Aalst, W.M.P., ter Hofstede, A.H.M., Weske, M.: Business process management: a survey. In: van der Aalst, W.M.P., Weske, M. (eds.) BPM 2003. LNCS, vol. 2678, pp. 1–12. Springer, Heidelberg (2003). https://doi.org/10.1007/3-540-44895-0_1
4. OMG: Business Process Model and Notation (BPMN) vol 2.0 (2011)
5. Chang, J.: BPM Systems: Strategy and Implementation. CRC Press (2016)
6. Dumas, M., van der Aalst, W., Hofstede, A.: Process-Aware Information Systems: Bridging People and Software Through Process Technology. Wiley (2005)
7. Delgado, A., Calegari, D., Milanese, P., Falcon, R., García, E.: A systematic approach for evaluating BPM systems: case studies on open source and proprietary tools. In: Damiani, E., Frati, F., Riehle, D., Wasserman, A.I. (eds.) OSS 2015. IAICT, vol. 451, pp. 81–90. Springer, Cham (2015). https://doi.org/10.1007/978-3-319-17837-0_8
8. Delgado, A., Calegari, D.: Systematic evaluation of business process management systems: a comprehensive approach. CLEI Electron. J. 21, 1–19 (2018)
9. Mendling, J., Reijers, H., van der Aalst, W.: Seven process modeling guidelines (7PMG). Inf. Softw. Technol. 52(2), 127–136 (2010)
10. Russell, N., Aalst, W.M.V.D., Hofstede, A.H.M.T.: Workflow Patterns: The Definitive Guide. MIT Press (2016)
11. Reijers, H., Liman Mansar, S.: Best practices in BP redesign: overview and qualitative evaluation of successful redesign heuristics. Omega 33(4), 283–306 (2005)

12. Neubauer, M.: E-Learning support for business process modeling: linking modeling language concepts to general modeling concepts and vice versa. In: Stary, C. (ed.) S-BPM ONE 2012. LNBIP, vol. 104, pp. 62–76. Springer, Heidelberg (2012). https://doi.org/10.1007/978-3-642-29133-3_5

13. Koutsopoulos, G., Bider, I.: Teaching and learning state-oriented business process modeling. experience report. In: Reinhartz-Berger, I., Gulden, J., Nurcan, S., Guédria, W., Bera, P. (eds.) BPMDS/EMMSAD -2017. LNBIP, vol. 287, pp. 171–185. Springer, Cham (2017). https://doi.org/10.1007/978-3-319-59466-8_11

14. Saraswat, S.P., Anderson, D.M., Chircu, A.M.: Teaching bpm with simulation in graduate business programs: an integrative approach. J. Inf. Syst. Educ. **25**(3), 221–232 (2014)

15. Recker, J., Rosemann, M.: Teaching business process modelling: experiences and recommendations. Commun. Assoc. Inf. Syst. **25**, 32 (2009)

16. Rudra, A., Jaeger, B., Aitken, A., Chang, V., Helgheim, B.: Virtual team role play using second life for teaching business process concepts. In: 2011 44th Hawaii International Conference on System Sciences, pp. 1–8 (2011)

17. Marjanovic, O., Bandara, W.: The current state of BPM education in Australia: teaching and research challenges. In: zur Muehlen, M., Su, J. (eds.) BPM 2010. LNBIP, vol. 66, pp. 775–789. Springer, Heidelberg (2011). https://doi.org/10.1007/978-3-642-20511-8_69

18. Dumas, M., Rosa, M.L., Mendling, J., Reijers, H.A.: Teaching BPM as MOOC. BPTrends (2015)

Digital Case Study: Augmented Process Analysis and Optimization

Wolfgang Groher[1]([⊠]) [iD] and Matthias Dietschweiler[2]

[1] Eastern Switzerland University of Applied Sciences, Rosenbergstrasse 59, 9001 St. Gallen, Switzerland
wolfgang.groher@ost.ch
[2] Sigvaris AG, Gröblistrasse 8, 9014 St. Gallen, Switzerland
matthias.dietschweiler@sigvaris.com

Abstract. We have developed a new approach for teaching business process analysis and optimization based on a digital case study. The case study has been elaborated in cooperation with a production company for compression wear, guided by the concept of action research and the ambition that students can virtually immerse themselves into the company's order-to-cash process. For this purpose, associated process documentation, stations of an order flow, interviews with employees and time stamps of process instances have been made available for students on the educational platform Moodle. The students cooperate in teams analyzing and optimizing the order-to-cash process with the objective to qualify for a consecutive consulting assignment. To gradually increase the level of complexity the students are guided through three phases of analysis building upon one another, being enabled by an interactive guide implemented with the Moodle plugin H5P. The fourth and final phase constitutes an open challenge in which the teams leverage all the information previously collected to identify optimization potential. They summarize their recommendations in a report addressed to a company representative. Feedback of students confirms a well-designed learning path as well as high practical relevance of this new education environment. The structure of the case applies action research at different difficulty levels and can directly be transferred to other companies regardless their business activity or industry. In a planned revision it will be examined how the Moodle plug-in "Level up" can be used to intensify the gamification aspect throughout the assignment.

Keywords: Order-to-cash · action research · augmented analysis · case study · problem-based learning · interactive

1 Initial Situation

Analyzing and optimizing business processes in a real environment requires profound skills in data collection techniques [4]. However, existing case studies are based on text descriptions and thus fall short in teaching data collection techniques, as all relevant data is already at hand. In addition, practical examples are greatly simplified for teaching in class with the result that assignments concentrate on modeling rather than on analyzing

J. Köpke et al. (Eds.): BPM 2023, LNBIP 491, pp. 222–229, 2023.
https://doi.org/10.1007/978-3-031-43433-4_15

and optimizing business processes. Subsequent analysis is limited because more in-depth information would have to be obtained for this.

Summarized, the key deficits in the status quo for case studies in business process management are:

- Unidimensional data basis: The case description focuses on textual information.
- Case information is already accessible: Learning the actual use of different data collection techniques is excluded from class teaching. Thus, time effort and informative value of different data collection techniques remain abstract for students.
- Artificial scenarios: Complexity of process analysis lags the reality of performing process analysis in a real-world environment.

2 Concept and Objectives

In practice, the analysis of business processes involves the combined application of different data collection techniques, such as study of process documents, observation, interviewing process employees and the quantitative evaluation of process data. Therefore, students need to immerse themselves into a real-world environment. They need to study process documentations and match this information with information provided by employees in interviews. In addition, they should walk along the single activities and tasks of a process and evaluate its performance based on quantitative data, such as lead times.

To consider these requirements, text-based case studies need to be expanded to include process documents, interviews with employees, video-sequences of the process flow and process time stamps. This can be enabled by converting the case study into a digital format to include all the additional information. The title "augmented process analysis and optimization" is derived from the fact that a text-based case study is extended by various real data and information to allow for an augmented view on the process. To enable students to learn from their own practice and experience, the approach follows the concept of action research as defined by Altrichter, Kemmis et al. [1].

The newly designed case study is aimed at students of business administration, business informatics and industrial engineering who are dealing with the topic of business process management. These students participate in an advanced level course (5th semester for full-time students, 7th semester for part-time students). The case is based on the Swiss production company Sigvaris, a manufacturer of compression products, which applies two different production strategies: make-to-stock and make-to-order. To limit complexity only the national, Swiss-internal business, is covered in the case. The following section addresses the didactic design of the case.

3 Learning Objectives

Derived from the teaching and learning situation as well as the above-mentioned concept, the following learning objectives can be formulated, grouped to the knowledge dimensions used by Conklin [3]:

- Conceptual knowledge
  The students know the common data collection and analysis techniques and their advantages and disadvantages.
- Procedural knowledge
  Students can structure the procedure for collecting process information as well as analyzing business processes and can apply the associated methods.
- Metacognitive knowledge
  Students can select and apply the most appropriate data collection and analysis techniques according to the individual situation.

4   Structure of Knowledge

In the case study there is increased emphasis on a procedural knowledge structure to be conveyed to students, because the focus is on the procedure for process analysis and optimization. This structure can be illustrated with the following four steps:

- Step-1: Understand the company business:
  Read and observe closely, based on a video on the stages of a customer order, digest the factual messages on business activities and the order-to-pay process.
- Step-2: Qualitatively analyze, understand, and model the order-to-pay process:
  Differentiate different types of orders: customer-specific products versus standard products, model associated processing steps, identify problem areas based on video interviews.
- Step-3: Quantitatively analyze, understand, and model the order-to-pay process.
  Identify and quantify differences between actual and target process e.g., lead time variances.
- Step-4: Identify opportunities for optimization
  Question and analyze deviations between actual and target process e.g., loops, setbacks, and delays and identify possible causes. Recommend "quick wins" and improvements that can be achieved in the medium term. Identify necessary information that must be provided by management as a prerequisite for further improvements.
  Objective and motivation for students in step-4 is to qualify for a consulting assignment.

5   Case Implementation

The implementation of the case study is guided by a problem based learning approach [6], which is, according to Engel, shaped by the following characteristics [5]:

- Cumulative – repeatedly reintroducing material at increasing depth
- Integrated – de-emphasizing separate subjects
- Progressive – developing as students adapt
- Consistent – supporting curricular aims through all its facets

With this as frame, the digital case study has been implemented on the learning platform Moodle, supported by the H5P-plugin, which permits interactive quizzes. This decision was taken, because Moodle is a standard platform in the Swiss university environment and can be used without additional license costs. From a technical point of view, any other platform that allows interactive quizzes would also be applicable. Processing of the case is divided into four sequences with the aim to guide students by gradually increasing the level of difficulty:

- Sequence-1: Getting to know the company and the stages of order processing.

     The students get an overview of business activities and environment of the company Sigvaris. In addition, a video is used to explain the run-through of a customer order. This simulates an on-site visit.
- Sequence-2: Focus on the order-to-pay process and its qualitative analysis.

     After the introduction, the students focus on an in-depth examination of the order-to-pay process. For this purpose, they have access to process artifacts e.g., process documents, measurement sheet for customized support stockings, and interviews with process employees documented as video sequences. To ensure that students grasp the key information, they are guided via an interactive quiz.
- Sequence-3: Quantitative analysis of the order-to-pay process

     Here, students concretize and supplement the results from the qualitative analysis with a process mining analysis, thereby obtaining real information, e.g., on lead times. Here, the students are also guided through an interactive quiz.
- Sequence-4: Assignment for group work

     In sequences 1–3 students have familiarized themselves with the initial situation at Sigvaris and gained an overview of the order-to-pay process. They now have a broad data and information base at their disposal:

   o Qualitative data/ information gained through

     - Process chain inspection
     - Process documentation
     - Interviews with process employees

   o Quantitative data/ information gained through

     - Process mining

Students are to use the knowledge previously acquired as base for the following and last part of the case study, which concentrates on the order-to-pay process with customers based in Switzerland. The final part of the case study is being worked on in groups of 3–4 students and results in a grade. The goal of each team is to qualify for a consulting assignment at Sigvaris. For this purpose, defined tasks have to be worked on and the proposed solution must be elaborated by each team. The corresponding document is then sent to Sigvaris as a project proposal. The document should have a size of four to six pages plus attachments. On this basis, Sigvaris management decides which team is awarded the consulting assignment.

The four sequences are summarized in Fig. 1, which includes the action research design principles applied to achieve competence to act with students.

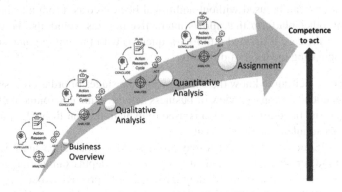

Fig. 1. Systematic capability building guided by action research

6   Implementation in the Classroom

The case has been piloted in two courses in fall 2022. The experience gained is reflected in the following section.

6.1   Structuring the Content

Since the analysis of the order-to-pay process is a complex task, processing of the digital case study should be done step by step. Accordingly, four steps are deployed:

Steps 1–3 serve to prepare students for the subsequent task in step-4 as well as to ensure that they are focusing on the appropriate section and instances of the process. This is secured by an interactive quiz, which was implemented with the Moodle plugin H5P.

Step-4 represents the actual test. Here, the students have to identify optimization levers in the process, based on the prior knowledge they have acquired.

The case study is implemented in Moodle in such a way that the next step in the case study is only activated for the students when a certain number of points (90% of the maximum possible points) has been achieved in the preceding quiz.

6.2   Time Structuring

A time span of three weeks is planned for the entire processing of the case. Students can use classes (2–4 lections per week) for working on the case. There is no teaching during this time, and the students are guided by interactive quizzes through the case. At the end of the three weeks, students must submit a report on identified optimization potential via Moodle.

6.3 Working in Groups

The case should be worked on in groups of 3–4 students. This considers the typical practical situation that a project team is usually used for such a task. In addition, this setting supports the discussion of the findings within the group and enables collective learning.

6.4 Guidance and Evaluation

In steps 1–3, students work independently in their groups, guided by the Moodle quiz, which ensures that they extract and interpret the essential information. During this time, the instructor is available each week for a one-hour time slot to clarify questions. This is to balance knowledge and experience differences across the different groups.

The report submitted in step-4 is evaluated based on a criteria set used to evaluate whether the students have been able to identify relevant cause-effect relationships and whether they have dealt with the case in such depth that they are also able to specifically name and demand missing information from the management.

In addition to this formal evaluation, a debriefing is carried out together with a company management representative in two lessons. The representative selects the soundest solution based on his practical perspective and awards it. This is supplemented by a discussion in which the student groups receive feedback on their solutions. These solutions are compared with actual decisions made within the company.

6.5 Avoiding a Transfer Trap

The goal of the case study is to make students recognize the strengths and weaknesses of different data collection methods using a concrete practical example. This is achieved implicitly through processing the case, since the students must use different data collection methods to answer the questions posed and compare their results.

The data collection methods chosen include studies of documents, process observation, interviews, and evaluation of process time stamps with the means of process mining. These methods are universally applicable, neither industry or business activity nor company size matter in this context. Therefore, the risk of a transfer trap is considered low in this setting.

Nevertheless, the debriefing should also include a sequence to discuss what differences might have to be considered when a service instead of a physical product is considered. With a service, the "production process" takes place in the interaction with the customers and cannot be recorded with time stamps of predefined activities, as it is the case in a customer advisory meeting, for example. In this case, increased attention must be given to interviews as data collection method.

7 Experiences from Piloting in Class

During the first implementation, an evaluation of the created teaching and learning material was conducted via a student survey. With respect a response rate of 24% it must be considered that the case study had been worked on in groups of 3–4 students.

Typically, the feedback was provided by the voice of one group so that effectively the overall response rate was greater than 75%.

7.1 Concept of the Digital Case Study

More than 55% of the responses rated the concept of a digital case study based on real company data as "very good", the remaining part of 45% as "rather good". Decreases in the rating result from the fact that there was no possibility to consult a company representative during the processing of the case study.

Lessons learned:

For clarification of questions, in future the lecturer will be available to the students for questions once a week for one hour. This is to balance differences in knowledge and experience across the different groups.

7.2 Achievement of Learning Objectives

The learning objectives defined were a higher transfer orientation than before and the creation of an environment in which students are "immersed" in the task ("immersive analytics"). More than 60% of the answers consider the achievement of this learning objective fulfilled within the range of "rather well" to "very well". Students partly complained difficulties in using the process mining tool Disco[1], which was previously introduced and used in class.

Lessons learned:

For clarification of questions regarding the operation of the process mining tool, the lecturer will be available to the students for questions once a week for one hour.

7.3 Design of the Quizzes/Learning Success Checks

The design of the quizzes/learning success controls regarding the aspects.

- Clear and understandable questions
- Verification of the solution
- User-friendliness

was rated "completely agree" or "rather agree" for more than 80% of the answers. Therefore, no adaptation is planned here.

7.4 Design of the Learning Path

The design of the learning path with the aim of a multi-level structure and a successive increase in complexity:

- Introduction to the problem on the basis of delimited assignments (drag & drop process modeling, multiple choice questions).
- Proposal for optimization (as proof of performance) in the form of open questions

More than 60% of the answers attested "completely agree" and more than 35% of the answers "rather agree". This approach will be maintained as originally designed.

---

[1] Disco is a process mining tool provided by the company Fluxicon BV.

7.5 Suggestions for Further Development of the Concept

The suggestions for the further development of the concept concentrate on the organizational aspect that it should be possible to ask questions to the lecturer during the processing of the case study. This will be realized in the form of a question-and-answer session once a week.

8 Conclusion and Outlook

For future implementations, the chosen approach with the current learning materials can be maintained. From an organizational point of view, a question-and-answer session for the students once a week will be offered to clarify content-related as well as methodological questions e.g., operation of the process mining tool Disco. In addition, it will be examined how the Moodle plug-in "Levelup" can be used to further implement the aspect of gamification for the assignment. What makes the case study stand out is its connection to current research in agile business process management [2]. There, people are looking at new technological approaches, such as process mining, with the aim of shortening the time required for process optimization.

Based on the experience gained with this new format of case study, the elaborated approach and structure can directly be transferred to other companies regardless their business activity or industry. However, making this transfer successful requires a close cooperation with the respective company as well as an open mind set on industry side. Access to the material created and used for the case study can be granted by the authors upon request.

References

1. Altrichter, H., Kemmis, S., McTaggart, R., Zuber-Skerritt, O.: The concept of action research. Learn. Organ. **9**(3), 125–131 (2002)
2. Badakhshan, P., Conboy, K., Grisold, T., vom Brocke, J.: Agile business process management. BPMJ **26**(6), 1505–1523 (2019)
3. Conklin, J.: A Taxonomy for Learning, Teaching, and Assessing: A Revision of Bloom's Taxonomy of Educational Objectives: Reviewed Work. Educational Horizons, pp. 154–159 (2005)
4. Dumas, M., La Rosa, M., Mendling, J., Reijers, H.A.: Grundlagen des Geschäftsprozessmanagements. Springer, Heidelberg (2021)
5. Engel, C.: Not just a method but a way of learning. In: Boud, D., Feletti, G. (Hrsg.) The Challenge of Problem-Based Learning. Kogan Page, London (1991)
6. Newman, M.J.: Problem based learning: an introduction and overview of the key features of the approach. J. Vet. Med. Educ. **32**(1), 12–20 (2005)

Developing Industry-Ready Business Analysts: Guiding Principles for Effective Assessment for Learning (AfL) in a BPM Course

Nina Evans⬤, Karamjit Kaur(✉)⬤, and Anisha Fernando⬤

STEM, University of South Australia, Mawson Lakes 5095, Australia
karamjit.kaur@unisa.edu.au

Abstract. This paper discusses a unique assessment approach that is implemented in the Business Process Management (BPM) course for Masters' students at the University of South Australia. The Active Learning and Assessment for Learning (AfL) principles are incorporated into this course to prepare industry-ready business analysts who will play an important role in the future business environment. We provide a detailed description of the design decisions, the implementation and resulting benefits of three major assignments in the course. The assessment is designed according to guiding principles that include 1) a focus on developing real-world skills for the future digital business environment, 2) following a process-oriented approach in both teaching and assessment, 3) motivating students to challenge assumptions and status quo by asking 'why' and 4) encouraging continuous self-reflection and a willingness to receive and provide constructive feedback. These guiding principles can be leveraged by other educators in the BPM community to design and implement an effective assessment structure for developing future-focused and industry-ready business analysts.

Keywords: Business Analyst · Business Process Management · Business Process Modelling · Active learning · Assessment for Learning

1 Introduction

Business analysts play a critical role by working with stakeholders to understand the organisation's current state, identify opportunities for improvement and define the business requirements for, amongst others, digital solutions. The education in Business Process Management (BPM) is crucial for the development of industry ready business analysts. Different educators employ different teaching and assessment techniques with varying degrees of effectiveness, when teaching a BPM course (Antonucci, 2010). This paper discusses how future business analysts' skills and attitude are developed through active learning and Assessment for Learning (AfL) approaches in the Masters' course Business Process Management (BPM) at the University of South Australia.

Related work about business process management education techniques found a lack of pedagogical resources and qualified instructors, large variations in content related to managerial and technical topics, limited pedagogical research on BPM education and

J. Köpke et al. (Eds.): BPM 2023, LNBIP 491, pp. 230–239, 2023.
https://doi.org/10.1007/978-3-031-43433-4_16

unclear BPM industry pathways for students (Bandara et al., 2010; Chakabuda, 2014). Further research identified a capability gap, related to industry demand in the Australian context (Mathiesen et al., 2013). To address the business process management education gap, the teaching and assessment design approach needs to be flexible to address these industry needs.

The teaching philosophy embodied in this course is active learning, which is based on "instructional activities involving students in doing things and thinking about what they are doing" (Bonwell & Eison, 1991) to empower students to acquire e.g. critical thinking, problem-solving, and effective communication skills (Brame, 2016). The active learning approaches (Michael & Modell, 2003; Prince, 2004; Michael, 2006; Cattaneo, 2017) as applied in this course are as follows:

- *Collaborative learning,* where students work together to solve problems, complete assessments, or critically discuss course material.
- *Inquiry-based learning,* as students are encouraged to listen and ask questions, explore new ideas, and seek answers through research and investigation.
- *Project-based learning,* where students interview a real-world client and apply BPMN skills to model as-is and to-be processes.
- *Case-based learning* involves students engaging in role-play and simulations to explore complex concepts or issues.
- *Problem-based learning* involves presenting students with case studies and challenging them to find solutions.
- *Discovery-based learning* involves students investigating their learning contexts through experimentation and trial-and-error based on specific requirements provided.

Active learning is done through a *flipped classroom* approach, where students are expected to read materials outside of class and use class time for discussions, problem-solving activities, or group work (Bishop & Verleger, 2013). Students are very actively involved in the learning activities and the assignments typically have high engagement and completion rates. Through active learning, students engage with new concepts, develop important skills, and reflect on their learning throughout the course (Michael & Modell, 2003).

In the next section we describe the unique way the assessment is designed for Learning and to develop the knowledge, skills and mindset attributes required of future business analysts.

2 Assessment Overview

In this section we describe the assessment philosophy and structure, emphasising the scaffolding aspects of the assessment design and highlighting the primary aims, tools used, and skills learned in each of the assignments.

2.1 Assessment Philosophy

Assessment philosophies focus on the central impact of student-teacher roles and relationships on the learning journey, a conducive learning environment and attaining specific educational outcomes (Schellekens et al., 2021). Assessment for Learning (AfL) is a suitable assessment approach in the context of business analysis, because it focuses on designing assignments that directly form part of the student's learning process and empower the learner through regular continuous feedback mechanisms (Wiliam, 2011; Brady et al., 2020). Instead of simply using assessment to record learning progress, AfL aligns teaching and learning strategies with assessment needs (such as industry relevancy) and furthers formative assessment strategies by identifying students' learning gaps and motivating them to engage in activities to address these learning gaps (Berry, 2008; Wiliam, 2011). In AfL, student learning needs are mapped to course learning outcomes, the instructional assessment design offers a reliable way to assess these needs, and student learning progress is regularly monitored (Berry, 2008).

Brady et al., (2020) propose five key AfL characteristics that can be embedded in course design, namely:

- *Balance of summative and formative assessment* where students experience a mix of assessed and non-assessed learning activities to encourage a more holistic reflection of knowledge and skills.
- *Authentic and complex assessment,* where the assessment design represents industry relevant contexts and higher-order learning skills.
- *Rich formal and informal feedback, as* a variety of quality feedback is provided at regular intervals through learning activities.
- *Opportunities for practice and confidence building,* as students grapple with new concepts or unlearn existing concepts before these concepts are formally assessed.
- *Opportunities to evaluate and direct own learning,* as students are encouraged to motivate and monitor their learning progress through formal and non-formal learning activities.

By embedding active learning and assessment for learning (AfL) design principles, students are less incentivised to seek the use of generative AI tools or other sources that risk academic integrity. Students are motivated to drive their own learning given the strong industry links and the opportunities to receive formative feedback on their learning progress.

2.2 Assessment Design and Structure

As shown in Fig. 1, the assessment of this course is designed and structured to develop a variety of skills through three scaffolded assignments. Assignment 1 builds skills required in assignment 2, which in turns builds skills that students will need in assignment 3. By scaffolding the assignments, we aim to support students to build confidence and become more independent learners, develop problem-solving skills to successfully complete complex tasks (Witt et al., 2019). Students are encouraged to continuously reflect on their learning and the processes used to complete each assignment. We ask questions that prompt students to think about what they have learned, what worked well, and what they might do differently next time. Students receive feedback on their work, and they are encouraged to use the feedback to improve their work.

By conducting a Systematic Literature Review, students develop skills in information searching, reading, analysing literature, and writing a literature review report. Students are assigned an industry and a list of research databases. The topic involves around

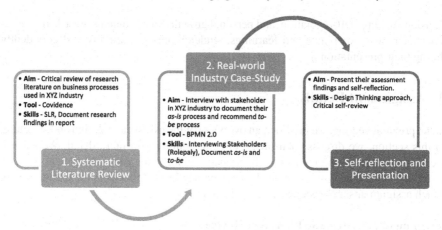

Fig. 1. Overview of the assessment structure

finding and documenting business process used in a specific industry in different contexts, for example how business processes have evolved through time or how they have changed with respect to digital business transformation within that industry. The process of finding and reviewing literature is not arbitrary if a Systematic Literature Review (SLR) methodology is followed. The web based Covidence tool is used to ensure that students follow the correct process for performing a systematic literature review. Finally, students produce a report about the business processes they found in their allocated industry. These insights are useful as a basis to conduct a personal interview with an external stakeholder from the same industry in the follow-up assignment.

The real-world case study allows students to develop professional communication skills with external stakeholders, to conduct a semi-structured interview, to analyse and model processes and to work effectively in a team. Students are randomly divided into groups of 3–4 to mimic collaboration in the real world. The group must find and interview a business analyst or manager of their choice in their assigned industry. The aim of the interview is to understand and document the *as-is* state of one business process. Students then analyse the business process and recommend an improved *to-be* state of the business process. The output report of Assignment 2 is shared with the industry stakeholder to elicit feedback about the *to-be* diagram and its implementation feasibility. In their course evaluation feedback, students often comment on both the professional and interpersonal skills they have learned as part of the groupwork and their interactions with a real-world industry stakeholder. The knowledge and skills developed in assignment 2 prepares the students for the next assignment.

In the assignment following the industry interview, students learn about doing an oral presentation, self-reflection, giving and receiving constructive feedback, and making recommendations to the industry client regarding process improvement. Self-reflection and dealing with feedback are important skills for students as it supports deeper learning, self-awareness, and growth. Self-reflection helps students develop a deeper understanding of themselves, their strengths, and areas for improvement. This can lead to better

decision-making, self-regulation, and personal growth. Since students are asked to articulate their own experience and learnings, students present naturally and confidently
during their presentation.

3 Assignment Details

In the previous section, we provided an overview of the assessment design and structure.
In this section, we discuss each of the three assignments in more detail. We explain
the aim of each assignment, how we prepare students for each assignment through
formative in-class activities, and the benefits gained by students from the unique way in
which assignments are designed and conducted.

Assignment 1: Systematic Literature Review
In Assignment 1, students conduct a Systematic Literature Review (SLR) to study the
business processes of a particular industry. The aim of conducting an exhaustive SLR is
to make students aware of the business processes that are executed in the industry and
appraise the technical jargon used in that industry.

To ensure that students execute the SLR process correctly, they must use the Covidence (https://www.covidence.org/) tool. Every step the students perform is documented
and the teaching team is added as reviewers in the setup of the Covidence account, which
provides the teaching team with a mechanism to audit the progress of students in the
various stages of SLR. Using Covidence to perform SLR ensures that students do not
use Artificial Intelligence tools (such as ChatGPT) to write the assignment.

Every student is assigned 3–4 specific research databases for their search. Both
academic and industry papers, including journal, conference, and white papers, can
be sourced. Students typically include more than 100 potential papers into Covidence,
before reviewing the abstract and eliminate the papers that are not completely relevant
for their topic. After reviewing the abstract, students must review the full paper in Covidence to identify the best 15–20 papers for inclusion in the final literature review report.
Students also create a data extraction spreadsheet with general information about the
papers, e.g., the publication type, publication year, keywords, research methodology, et
cetera. We ensure that students perform the process steps within the timeline, by including regular milestones. For the literature review, students highlight the most important
findings of previous research, they compare different viewpoints, and write an integrated
and concise report that includes quotes, paraphrasing and in-text referencing.

To equip students with skills to execute the SLR, detailed documentation is provided
including suggestions on use of appropriate combinations of search keywords for the
best search results. Dedicated workshops are held to walk through an example SLR
process and to assist students with the use of the Covidence tool to complete the SLR
process.

Assignment 2: Real-World Industry Case Study
This assignment prepares students for the capstone project course and a career as a business analyst. Students receive real-world exposure when they conduct a personal interview with an industry stakeholder to understand one of the existing business processes in

their organisation and identify ways to improve the process. They learn to communicate professionally when trying to find a suitable interviewee (through personal contacts, email, or LinkedIn) and during the interview.

The knowledge about the allocated industry sector, as obtained in the SLR, is leveraged in Assignment 2. This is a group assignment and hence students learn skills to work collaboratively in pre-defined groups to find and interview an industry stakeholder. They develop and document the interview questions beforehand and aim to elicit as much information as possible about a particular business process used in the interviewee's organisation. Students use the BPMN 2.0 notation to create the process diagram to model the *as-is* and *to-be* processes of the organisation.

The report produced by the students is shared with the industry partners to receive feedback about the *as-is* diagram and the feasibility of the suggested improvements as captured in the *to-be* improvement diagram.

To prepare the students for this assessment, we simulate a 'ball in the bucket' process where students try to toss a ball in a bucket from a distance. Through playing the game students learn to identify and improve process inefficiencies. In preparation for the real-life client interview, students experience a mock user requirements elicitation during a role-play exercise, to develop empathy, critical thinking, and problem-solving skills. In addition, guest lecturers are invited to discuss the role of the business analyst in industry and address students' questions. During workshops, students practice the drawing of the *as-is* and *to-be* states of business processes in BPMN 2.0 notation based on multiple case studies of varying complexity.

Assignment 3: Self-reflection and Presentation

In this assignment students reflect upon their strategy and approach to the previous assignments (i.e., the process they used) and the outcomes they achieved. While doing so, students must consider the design thinking approach and how it overlaps with, and differs from, the BPM lifecycle approach they followed in Assignment 1 and 2, While developing a *to-be* state of business processes, students often do not consider the feasibility of implementing and testing their suggested solutions. In Assignment 3, we ask students to brainstorm and deliver an oral presentation about their findings from Assignment 1 and 2, to discuss the implementation of their suggested improved *to-be* business process and the lessons learned. They also share the feedback received from their industry stakeholder and many of the student groups realise that they did not ideate well enough, and that the changes they suggested are not realistic to implement, especially from the time and budget constraints point of view.

We can overlay the design thinking principles with the three major assignments. Assignment 1 corresponds to the *Empathise* stage, where students become aware of the processes used in the industry through conducting an SLR. Assessment 2 corresponds to the *Define* and *Ideate* stage where students *empathise* with the industry stakeholders by understanding the existing *as-is* business processes used in the organisation. After the interview, students *ideate* to discuss possible solutions to improve the *as-is* business process and suggest the best solution as the *to-be* process. As preparation for assignment 3, students are expected to regularly reflect articulate their viewpoints in tutorial sessions. The design thinking approach and its underlying principles are explained to students in detail.

The next section describes the guiding principles underpinning the assignments in this BPM course. Other educators in the BPM community can benefit from incorporating these principles in their course assessment.

4 Guiding Principles

In all the BPM assignments (as explained above), we follow four underlying guiding principles (GP1-GP4), which are explained in this section. We have explained our approach in this paper, which can be adapted as required. We recommend that the BPM teaching community incorporate these principles in the design of their assessment structure. This can be done in various ways.

Below is a summary of the principles we follow in this course:

GP1: Cultivate real-world skills to produce industry-ready business analysts
The course develops an understanding of a specific industry and the processes it employs, through conducting an SLR. Students are given an opportunity to communicate with a real-world industry stakeholder by connecting them professionally, interviewing them, and asking for feedback about real world implementation constraints (such as time and budget). The seminars in this course include discussions about real-world organisations such as Toyota and Uber, and real-world challenges (e.g., disruptive technologies and COVID-19) and business process improvement approaches (Kaizen and Six Sigma).

The course also provides graduates with a foundation towards preparing for industry certification from e.g., the International Institute of Business Analysis (IIBA) or the Project Management Institute (PMI).

GP2: Follow a process-oriented approach
Understanding the importance of processes and following a process to execute a task are very essential skills for a graduate to be ready for industry (Hrabala et al., 2017; Seethamraju et al., 2012). A process approach is followed throughout the teaching and assessment of the course. Students apply the process embedded in the Covidence tool to conduct a systematic literature review, which illustrates the advantages of executing a process in steps. Students are exposed to both business process management lifecycle and design thinking, and how these two approaches differ.

GP3: Challenge assumptions by asking *why*
The BPM whole course is based on the principle that changes/improvements to business processes are not made for the sake of making them, but for a specific reason and in line with a business' strategy. Asking "why" is a fundamental aspect of a business analyst's critical thinking, problem-solving, and decision-making. It is important because it helps students to understand the root causes of a situation or problem, and to identify the underlying motivations, assumptions, and beliefs that are driving their own or others' behaviours. Asking "why" is a powerful tool for gaining deeper insights, challenging assumptions, and making better decisions, whether one is working on a complex problem or simply trying to understand a situation. By asking "why" repeatedly, students can drill down to the underlying cause of an issue rather than simply treating the symptoms. We

find that students from countries with a high 'power difference' (Hofstede, 2011) are not encouraged to challenge and ask why and this course provides them the freedom to do so in a safe learning. As future BAs their mindset changes to be more self-directed, motivated, and capable.

GP4: Encourage self-reflection and willingness to receive/provide constructive feedback

Self-reflection and giving/receiving feedback are important skills for students as they support deeper learning, self-awareness, motivation, and growth (Jalali, 2018). In this course a whole assignment is dedicated to self-reflection and feedback. Students are encouraged to think about what they've learnt. They receive feedback from other students, lecturers, and the industry stakeholder. Students also provide feedback to each other and to lecturers about ways to improve the course. The oral presentation builds students' confidence in articulating their thoughts, because they're talking about their own experience and mistakes they've made while completing the assignments.

5 Discussion

The overarching goal of this course is to prepare students for the capstone project and for the world of work, by providing then with useful and applicable business analysis skills and knowledge. In this paper we suggest the key guiding principles for assessment design that we apply in this BPM course, which integrate active learning and assessment for learning characteristics.

Active learning (as a teaching and learning philosophy) and AfL (as an assessment philosophy) are both useful approaches in this BPM course because both philosophies empower students to accomplish learning goals that address BPM future industry needs, through scaffolded feedback-driven activities and regular assessment where students are invited to learn, unlearn and relearn BPM concepts and skills. In this course we scaffold the assignments to build upon previous learning. We provide exposure to real-world problems, follow a process-oriented approach in teaching and assessment, and encourage self-reflection and feedback. Feedback can help students understand their progress and provide a sense of achievement and it can also help students identify areas for improvement in their work, leading to higher quality outputs and a deeper understanding of the subject matter. This can increase motivation and engagement in the learning process. Giving and receiving feedback also helps students develop communication skills as they learn to articulate their ideas, accept criticism, and provide constructive feedback to others.

Another important principle of the BPM course is to teach students to ask *why*, instead of blindly accepting what they are taught or what the user asks of them. Student and industry feedback provides a useful perspective on the BPM assessment design principles in this course.

These guiding principles for designing assessment components are mapped against relevant active learning approaches and AfL characteristics for each assignment, as summarised in Table 1.

Table 1. BPM Course Assignment Design Principles Mapped Against Active Learning Approaches and AfL Characteristics

Assignment	Guiding Principle	Active Learning Approach	AfL Characteristics
Assignment 1	GP1, GP2, GP3	Inquiry-based learning, Problem-based learning, Discovery-based learning	Authentic and complex assessment, Rich formal and informal feedback
Assignment 2	GP1, GP2, GP3	Collaborative learning, Inquiry-based learning, Project-based learning, Case-based learning, Problem-based learning	Authentic and complex assessment, Rich formal and informal feedback, Opportunities for practice and confidence building, Chances to evaluate and direct own learning
Assignment 3	GP1, GP2, GP4	Collaborative learning, Inquiry-based learning, Project-based learning, Discovery-based learning	Authentic and complex assessment, Rich formal and informal feedback, Opportunities for practice and confidence building, Chances to evaluate and direct own learning;

6   Future Research

Due to the limitation on the length of this paper, we plan to include further details in an extended paper about the effectiveness of the assessment structure and the guiding principles explained in this paper. This evaluation will be based upon the quantitative and qualitative feedback received from the students at regular intervals and at the end of the course. Additionally, we will provide details on how the learning outcomes from the BPM course connects with, and benefits other courses being taught in the Masters' program at the University of South Australia.

We also plan to provide further details about how the BPM course and other courses in the program prepares our graduates for their important role in the Digital Business Transformation of an organisation and to deliver 'Enterprise 4.0-ready business analysts'.

References

Antonucci, Y.L.: Business process management curriculum. In: Handbook on Business Process Management 2: Strategic Alignment, Governance, People and Culture, pp. 423–442 (2010)

Bandara, W., et al.: Business process management education in academia: Status, challenges, and recommendations. Commun. Assoc. Inf. Syst. **27**(1), 743–776 (2010)

Berry, R.: Assessment for Learning. Hong Kong University Press (2008). http://www.jstor.org/stable/j.ctt1xcs68

Bishop, J., Verleger, M.A.: The flipped classroom: a survey of the research. Paper Presented at 2013 ASEE Annual Conference & Exposition, Atlanta, Georgia, June 2013. https://doi.org/10.18260/1-2-22585

Brame, C.: Active Learning. Vanderbilt University Center for Teaching (2016). https://cft.vanderbilt.edu/active-learning/

Brady, M., Fellenz, M.R., Devitt, A.: Adopting assessment for learning (AfL) in higher education: implications for technology deployment. In: Allen, S., Gower, K., Allen, D.K. (eds.) Handbook of Teaching with Technology in Management, Leadership, and Business, pp. 418–432. Edward Elgar Publishing, Cheltenham (2020). Chap. 38. https://doi.org/10.4337/9781789901658

Bonwell, C.C., Eison, J.A.: Active learning: creating excitement in the classroom. ASH#-ERIC Higher Education Report No. 1. The George Washington University, School of Education and Human Development, Washington, D.C. (1991)

Cattaneo, H.: Telling active learning pedagogies apart: from theory to practice. J. New Approaches Educ. Res. 6(1), 144–152 (2017). https://doi.org/10.7821/naer.2017.7.237

Chakabuda, T.C., Seymour, L.F., Van Der Merwe, F.I.: Uncovering the competency gap of students employed in business process analyst roles—an employer perspective. In: 2014 IST-Africa Conference Proceedings, pp. 1–9. IEEE (2014)

Hofstede, G.: Dimensionalizing cultures: the hofstede model in context. Online Readings Psychol. Cult. 2(1) (2011). https://doi.org/10.9707/2307-0919.1014

Hrabala, M., Opletalováb, M., Tučekc, D.: Teaching business process management: Improving the process of process modelling course. J. Appl. Eng. Sci. 15(2), 113–121 (2017)

Jalali, A.: Teaching business process development through experience-based learning and agile principle. In: Perspectives in Business Informatics Research: Proceedings of the 17th International Conference, BIR 2018, Stockholm, Sweden, 24–26 September 2018, vol. 17, pp. 250–265. Springer International Publishing (2018)

Mathiesen, P., Bandara, W., Marjanovic, O., Delavari, H.: A critical analysis of business process management education and alignment with industry demand: an australian perspective. Commun. Assoc. Inf. Syst. 33(27), 463–484 (2013). https://doi.org/10.17705/1CAIS.03327

Michael, J.A., Modell, H.I.: Active Learning in Secondary and College Science Classrooms: A Working Model of Helping the Learning to Learn. Erlbaum, Mahwah (2003)

Michael, J.: Where's the evidence that active learning works? Adv. Physiol. Educ. 30, 159–167 (2006). https://doi.org/10.1152/advan.00053.2006

Prince, M.: Does active learning work? A review of the research. J. Eng. Educ. 93(3), 223–231 (2004). https://doi.org/10.1002/j.2168-9830.2004.tb00809

Schellekens, L.H., Bok, H.G., de Jong, L.H., van der Schaaf, M.F., Kremer, W.D., van der Vleuten, C.P.: A scoping review on the notions of Assessment as Learning (AaL), Assessment for Learning (AfL), and Assessment of Learning (AoL). Stud. Educ. Eval. 71(101094), 1–15 (2021)

Seethamraju, R.: Business process management: a missing link in business education. Bus. Process. Manag. J. 18(3), 532–547 (2012)

Wiliam, D.: What is assessment for learning? Stud. Educ. Eval. 37(1), 3–14 (2011)

Witt, C.M., Sandoe, K., Dunlap, J.C., Leon, K.: Exploring MBA student perceptions of their preparation and readiness for the profession after completing real-world industry projects. Int. J. Manage. Educ. 17(2), 214–225 (2019)

Managing BPM Projects and Their Implications on Organizations – Experiences from Research and Teaching

Manuel Weber[(✉)] [iD] and Jan vom Brocke [iD]

Liechtenstein Business School, University of Liechtenstein, Vaduz, Liechtenstein
{manuel.weber,jan.vom.brocke}@uni.li

Abstract. Business Process Management (BPM) is a holistic management discipline in which numerous facets and dimensions can be addressed within education and teaching. Especially for novices (in educational institutions and practice), the large body of knowledge can be perceived as overwhelming due to the pluralistic background and its resemblance to other related management disciplines. Also, especially in real-world settings, BPM initiatives must be considered holistically since local process improvements rarely yield satisfactory results. Within this contribution, we report on our experiences using the BPM Billboard in several practitioner workshops or student seminars in which we regularly utilize this tool to discuss or showcase BPM initiatives on a high- and holistic level. We apply the BPM Billboard to an ongoing real-world project in the context of e-government and discuss the implications for the organization. The results indicate that practitioners and students need to master this basic knowledge to successfully communicate and coordinate large-scale projects across the organization.

Keywords: Business Process Management (BPM) · BPM Billboard · E-Government

1 Introduction and Motivation

Understanding and managing change in organizations has become one of the most critical success factors for maintaining organizational performance and competitiveness. Therefore, training and education initiatives are even more important for interns and students who will later have to compete in the industry. To successfully seize and scope large-scale projects in the corporate context, practitioners can utilize various models and frameworks as found in handbooks or scientific articles. However, few tools or approaches go beyond traditional process thinking and process modeling [1, 2].

Within this contribution, we report on our experiences at the interface of teaching and research in BPM and point to the importance of knowledge transfer to students and practitioners. We report on the use and benefits using the BPM Billboard by vom Brocke et al. [3], which we regularly use in workshops with representatives from several industries, and students from our Information System (IS) Master's class. Based on scientific

© The Author(s), under exclusive license to Springer Nature Switzerland AG 2023
J. Köpke et al. (Eds.): BPM 2023, LNBIP 491, pp. 240–249, 2023.
https://doi.org/10.1007/978-3-031-43433-4_17

results and grounded in theory, such a model not only facilitates discussions within a student's class or practitioner's workshops to showcase their projects but also proves useful in highlighting key factors relevant to manage projects holistically and successfully. Moreover, by applying this holistic approach, novices can work out the basics of BPM as a process-oriented management discipline and gain valuable insights into the interplay between BPM and the organization, its structure, and strategy. To showcase this statement, we describe the real-world case of a State Department in the context of electronic government (e-government) located in Central Europe. The State Department is undergoing a significant wave of digitalization initiatives and is confronted with adopting new information technologies (IT) to realize the re-design of its core business processes. The case results demonstrate that such a tool captures the key elements in BPM and reveals the impact and associated changes on the organization.

The remainder of this contribution is structured as follows: After this brief introduction (Sect. 1), we report on the problem and issue being addressed (Sect. 2). In Sect. 3, we briefly report on the goal of this paper and present the results in Sect. 4. Finally, Sect. 5 presents the benefits for the BPM educators community and beyond.

2 Problem and Issue Being Addressed

As part of our research activities in Business Process Management (BPM) and Information Systems Research (ISR), we organize workshops and regularly hold seminars and lectures at various universities. We encounter experienced managers and decision-makers from SMEs or local authorities and first-semester students who are required to describe and analyze BPM initiatives as part of their seminar, bachelor, or master theses.

The core idea of these seminars is the interlinking of science and practice and, thus, the transfer of scientific and practical results. We present the latest concepts from BPM science and also offer a kind of "reality check" for the participants. Hence, we provide content from our research or collaborations with other companies and deliberately incorporate a "hands-on" element by asking participants to analyze and present their projects from their business environment or trainees to the other participants. In these "learning by doing" sessions, participants from different industries and knowledge bases can learn from each other. Valuable discussions arise, leading us as researchers to new insights and approaches. Finally, according to the motto "show, don't tell", we want to get them to report about their pitfalls and success stories.

Practitioners and managers consider BPM to be a panacea and a goal itself. However, research has shown that BPM should never be pursued as a goal but rather as a means. Accordingly, we see BPM as a driver or enabler of digitalization projects and the implementation of corporate goals [4]. The same applies to new information technologies (IT), which are closer to the process than ever before. For example, process mining enables the investigation of process behavior by analyzing process logs, the digital traces left by each information system, and process stakeholders along a digitized process [5].

However, based on our experiences from university-industry collaborations and research projects, we learned that fundamental and overarching issues such as strategy and governance are often neglected and rarely considered when re-designing single processes. Therefore, we often miss the socio-technical view of processes in practice and with students, coupled with divergent discussions and views on different altitudes. Particularly in a multi-dimensional setting such as the ongoing digital transformation in which most of our partner organizations find themselves, process stakeholders must communicate on the same level and solve complex problems. One-sided and isolated considerations are a central challenge in today's dynamic environment. Therefore, organizational change can occur on many levels [7, 8] and must be structured and consistent with the organization's strategy and the prevailing BPM paradigm.

Among other issues, this myopia partially originates from the pluralistic background of BPM as a process-oriented management discipline and a key concept in information systems research (ISR) [6, 7], but also due to the resemblance with other management and research disciplines such as information systems research or organization science, in which processes often the main focus. Although we do not expect practitioners, early-career researchers, or even students to be familiar with the extensive literature, we argue that BPM education and training beneficiaries must be equipped to deal with complex issues in practice and have a sound basic understanding of multi-dimensional challenges. Finally, as an advanced BPM (education and even research) community, we must attempt to successfully communicate the special aspects and particularities of our process-oriented management discipline and advertise for it.

3   Goal of the Contribution

Within this contribution, we draw attention to the abovementioned topics and share our experiences from being involved in practice-oriented research projects and teaching in different educational institutions and forms. In addition, we contribute to findings in the literature to make BPM training and education more efficient for the participants and to provide them with added value [2]. To this end, we present a tool that has received little attention in research and practice. Its application proves intuitive and is therefore suitable for various teaching purposes, as it considers high-level elements and capabilities relevant to successfully conducting BPM. Presenting a case from a real-world and ongoing project in a State Department of Economic Affairs, we demonstrate the tool's suitability for developing information systems (IS) and re-designing processes. This procedure aligns with research findings demonstrating that contemporary BPM goes beyond individual aspects of methods or tools and instead addresses the management of the associated capabilities [7]. Therefore, we offer a way to promote BPM as a modern and competitive discipline and make it more accessible to novices and practitioners. Finally, applying the tool to the case, we demonstrate how to use the tool to get practitioners and students to engage in fundamental discussions at a common level and consider the integrative nature of BPM within the organizational context.

In the following paragraphs, we first describe (and recap) the BPM Billboard by vom Brocke et al. [3] and then use it to present the case organization and one of its ongoing projects.

4 Analyzing a Real-World Project Using the BPM Billboard

Within the BPM and IS research community, numerous models and concepts have been developed to conceptualize and elaborate on projects or re-design initiatives within organizations. In addition to reducing complexity, the aim is to make abstract content tangible and present it in a comprehensible way [e.g., 8, 9]. This tool is not only available in published version [3] but can also be accessed via the website (https://www.bpm-billbo ard.com). The contact details of the authors are also available, who support practitioners in their projects through their many years of experience and industry contacts.

The BPM Billboard, designed as a one-page illustration (see Fig. 1), intends to support practitioners in linking their projects to the corporate strategy and assess the organizational capabilities relevant to its realization. Moreover, the framework is easy to understand and especially valuable for practitioners to convey the essential elements from BPM research to practice. Next, the BPM Billboard has been published in a leading BPM handbook dedicated to showcasing contemporary and real-world case studies for practitioners, scholars, and educators by applying a holistic approach [10]. Besides its practice-oriented approach, the elements of the BPM Billboard are empirically grounded, underlying the theory and research about maturity models. In addition, it has been empirically validated and refined [4].

It also proves useful for establishing the (allegedly) missing link between business, IT, and strategy [1] – in addition, far more elements are considered (governance, method, people, culture, etc.). Moreover, it proves useful to highlight key dimensions in BPM as it covers the six core elements and the associated capability areas, which have gained popularity in BPM research and beyond [4, 7]. In addition, we find the strategic and operational levels (represented by the strategy, strategic alignment, and project results) relevant to understand the implications and change in organizations [11, 12].

In previous studies, we also find evidence to use such a (maturity) model to study organizational change. Following the study of Andreasen and Gammelgaard [13], we find attempts to guide the investigation of change using maturity models – the underlying idea of six core elements within the BPM Billboard. Along the same line, Andersen and Henriksen [14] also propose a maturity model to discuss the caused changes triggered by digitalization initiatives. Finally, we argue that applying the framework contributes to examining change in BPM and the respective organization.

Our case organization is a State Department located in Central Europe. It is generally responsible for supervising and implementing the government's economic policy. Among other areas, the State Department supervises trade law. As part of this sovereign task, they verify and approve trade licenses for corporations that conduct business in the country or which offer inbound commercial services. As of today, the State Department identified 32 trade law processes, which they visualized using a Business Process Modeling Notation (BPMN). In 2019, the State Department launched a project to develop a new IS platform for re-designing and managing their trade law processes end-to-end. In 2022, the Parliament passed a legal basis for implementing electronic communication between authorities, citizens, and local companies (their "clients"). The idea is that citizens and corporations exchange data and documents with the State Department solely electronically using the new "e-government IT services" (in short, "e-gov services").

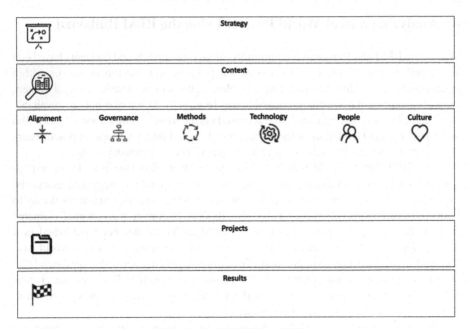

Fig. 1. The BPM Billboard [3].

The State Department's current information system (IS) – we refer to it as the "Trade Register Application (TRA)" is outdated from a technical point of view and thus has reached end-of-life. By now, it has served as a simple register to store metadata of trade licenses. To apply for an extended trade license, customers had to fill out PDF forms and send them to the State Department. The data transfer into the TRA was done manually by the officers. To coordinate work between the officers, they used an Excel file to store and highlight incoming customer requests. In the past, customers could not check the current processing status online but had to call the State Department by phone or contact it via email. In a nutshell, the work for the officers was time-consuming and cumbersome (Table 1).

Using the BPM Billboard, we identified five paradigm shifts the case organization is currently facing. Subsequently, we briefly describe the upcoming changes and paradigm shifts and link them back to the elements of the BPM Billboard (see italic and *CAPITALIZED* elements in brackets). These effects show the new understanding of end-to-end processes within the case organization and reflect the specific impact of such a complex project on the entire organization. With these tangible links, such multi-layered initiatives can be better communicated within an organization:

Paradigm shift no. 1 refers to the closer customer integration into the supervisory processes. Before the new IS platform replaced the Trade Register Application, customers had no transparent overview of pending applications and trade licenses. Also, the customers had to use other communication channels, such as telephone or email, to check the status of their application and ask for details. Now, the IS platform provides the customer with an overview of the status and details of the customer's applications

Table 1. Analyzing the Project Using the BPM Billboard, adapted from [3].

Strategy (of the project)	The new IS platform and processes will be re-designed from an end-to-end perspective. The State Department and its customers can communicate with each other without media discontinuity and process applications for trade licenses electronically and efficiently. Integrating the "e-gov services" results in a modern and efficient platform for managing and registering trade licenses
Context	The processes and tasks of the State Department are determined by several Acts and Ordinances. In e-government, these regulatory adjustments considerably influence the re-design of processes. Within these regulations, however, there is no process orientation explicitly recognizable. Nevertheless, one can find strict requirements and checks and balances among the several actors mentioned. At the very least, this enables the process-related distribution of roles and responsibilities, which in turn influence the design of the new IS platform
(Strategic) Alignment	The development of the IS platform implements the government's requirements and digitization strategy. Corporations and the State Department must communicate with each other exclusively electronically
Governance	Previously, responsibilities within the State Department were distributed among several specialists, and each officer acted within the scope of their professional expertise. Today, they can independently assign orders from a central order book. Responsibilities are distributed. This shared knowledge enables better deputy regulation within the State Department
Methods	The new IS platform (a web application) re-uses relevant customer data multiple times. The methods for retrieving and processing information have changed fundamentally ("once-only principle"). In addition, there is a stronger focus on the customer and his integration into the processes ("push and pull principle") A multi-view is applied to the processes to better manage the requirements and goals of the process stakeholders involved. Customers can view existing trade licenses more transparently in their self-service area and modify them (with subsequent approval by the officers)
(Information) Technology	The e-government IT services relevant for the identification, representation, payment, and delivery ("basic functions") significantly determine and influence the re-design and realization of the trade law processes
People	Using the new IS platform requires new skills to view processes from an end-to-end perspective. Due to the high level of customer integration, employees must demonstrate a high degree of understanding of the entire process landscape

(continued)

Table 1. (*continued*)

Culture	The mode of operation and handling of digital files leads to a new way of working and a new work culture. Digital files also make working in the home office or on business trips possible. Physical files were usually not allowed to leave the workplace and had to remain in the State Department due to security standards
Projects	The project team depends on the developments and advancements of e-government IT services. Renewals must be constantly reviewed and implemented within the framework of the new IS platform
Results	The result of the project is that commercial law topics can be mapped end-to-end and digitally. The project leaders see further potential for mapping the IS platform around other topics and for other areas of law

and requests. Customers can access and log in to the new IS platform via a web service portal using their "e-identification". This shift is not only based on the strategic viewpoint (*STRATEGIC ALIGNMENT*) of redesigning and managing processes for enhanced customer integration but also leads to changed process descriptions and visualizations (*METHOD*). Due to the enhanced customer integration and their extended possibilities to intervene in the business processes, the State Department officers need to evolve their skills in thinking across their process landscape and understand the interdependencies (*PEOPLE*).

In **paradigm shift no. 2**, the triggering of processes is no longer unilateral and distributed over several channels. Rather it is bi-directional in the sense of a push and pull principle: Until now, customers had to find a workaround (phone or email) to contact the relevant State Department officers to retrieve the status of their applications or requests. Now, the IS platform enables triggering services and processes by the customer or the State Department through a self-service approach. For example, ordering an extract from the register will be done without the intervention of the State Department officers. This "four-eyes principle" and role is omitted here (*GOVERNANCE*). Rather the customers can order such an extract on their own. Moreover, customers will automatically receive notifications via email or within the self-service area. Also, the customer can withdraw initiated processes. This principle will revolutionize the way the two parties communicate with each other. Both parties can initiate processes and services via the "push principle" or request them via the "pull principle. What changes is the distribution of roles and rights. Although approvals are still reserved for the State Department (as described by law), customers can now view their data in real time and initiate adjustments on their own (*GOVERNANCE*).

We also identify **paradigm shift no. 3:** The obligation to communicate (exclusively) electronically with corporations and the adoption and use of e-gov services results in a further paradigm shift. According to law, a mandatory requirement is to re-use data the customers had entered once (*legal CONTEXT*). Until now, data provided by applicants or generated by the State Department's officers had to be entered multiple times across the processes. Now, data can be exchanged more easily across other State Departments.

We identified **paradigm shift no. 4:** The customer's file and records, including the application forms and attached documents, were previously kept as a physical file. Customers could upload the attachments while filling out the online form. A web service then generated PDF files, which had to be manually downloaded by the Department officers. It was also possible for the customers to send the documents via the postal service. The physical documents were scanned and stored as PDF files in a specific file storage. The E-Government Act stipulates that documents and the customer's files may only be kept electronically (*LEGAL CONTEXT*). Digital documents will become the so-called "digital primacy" at the moment of introduction. As soon as documents are scanned and transferred to a document management system (DMS) (*IT*) via the interface, they are legally perceived as the "original documents". Furthermore, the DMS allows to electronically sign documents.

Paradigm shift no. 5 represents the enhanced alignment and integration of business and organizational processes (*STRATEGIC ALIGNMENT*). For years, the State Department officers had different priorities in processing applications. To coordinate the incoming applications, one officer is assigned to coordinate the processing of the tasks and clarify any legal issues. The assigned officer had to ender the applications into an Excel list. With the new IS platform, an "order book" was created to map all customers' applications (new trade licenses or adjustments to existing ones). Consequently, every officer has the opportunity to assign orders independently and manage the application process end-to-end. Internally, the State Department can easily assign the responsibilities of received applications to other employees and automatically trigger follow-up processes (*GOVERNANCE*).

5   Impact and Benefits for the BPM Educators Community

Based on our experiences from research and teaching, we were able to derive three main and general lessons learned:

The systematic capturing of key factors in BPM should lead to a holistic picture of how processes are managed and integrated into the organizational context. In this regard, considering the organizational context enables educators to teach BPM in an integrated way and thus not isolated from other topics. The elements covered are based on research findings and grounded in maturity model theory and a context-aware management approach [e.g., 15]. Therefore, we advocate that the BPM educators community increase the use of concepts, models, and tools supported by theories from organization science or management research that enable the connection between theory and practice. The advantages are obvious: Research can reflect valuable insights and findings from practice and examine them scientifically. This approach enables us to reduce research latency and the gap to rapidly integrating and publishing results valuable to the community. In this context, we still see the potential for the BPM community to move closer to practice and vice versa. To meet this demand, the annual International Conference on Business Process Management offers professional and promising platforms for researchers and practitioners. Finally, we want to encourage the BPM (education) community to develop further tools and platforms that move research and practice closer to each other.

Moreover, such a framework should not offer a one-size-fits-all solution but empower the participants and allow them to work out the essential factors in a context-sensitive way. The BPM Billboard meets this requirement as it not only demands and considers the respective context but also includes the organizational strategy and, thus, the realization of the corporate goals. We learned that such a bottom-up approach also triggers the learning process and enables participants to develop "their own projects". In addition, practitioners reported that with top-down management approaches, the (top) executives in charge may be unfamiliar with all the issues in detail (on operational level) and therefore have little knowledge of what is actually happening.

Applying the BPM Billboard, we demonstrated that BPM should not be viewed in isolation from the organizational context. Rather, it represents a framework that requires integration into organizational structures, strategy, and tangible outputs. In this regard, it provides a basis for discussion so that any new project can be holistically aligned with the organization, which ultimately releases and provides the resources for BPM [15]. This increased understanding and knowledge of which key elements and capability areas of the organization are critical, enables participants to achieve a greater emphasis on the communication and coordination of projects across the whole organization. This asset or skill enables them to better represent their expertise to the management and beyond. Ultimately, getting the key stakeholders and decision-makers involved is a matter. They must be convinced of new projects and support the decisions. Ideally, this will lead to less resistance within the organization and its stakeholders.

Finally, we mention that such tools can never fully cover all conceivable factors or needs of project stakeholders. We would like to point out that no model guarantees that projects will be completed on time or even in the most efficient way. Nevertheless, they can help to conceptualize a certain aspect or level of consideration.

References

1. Seethamraju, R.: Business process management: a missing link in business education. Bus. Process. Manag. J. **18**, 532–547 (2012)
2. Thennakoon, D., Bandara, W., French, E., Mathiesen, P.: What do we know about business process management training? Current status of related research and a way forward. BPMJ **24**, 478–500 (2018)
3. vom Brocke, J., Mendling, J., Rosemann, M.: Planning and scoping business process management with the BPM Billboard. In: vom Brocke, J., Mendling, J., Rosemann, M. (eds.) Business Process Management Cases Vol. 2, pp. 3–16. Springer, Berlin, Heidelberg (2021). https://doi.org/10.1007/978-3-662-63047-1_1
4. Rosemann, M., vom Brocke, J.: The Six Core Elements of Business Process Management. In: vom Brocke, J., Rosemann, M. (eds.) Handbook on Business Process Management 1, pp. 107–122. Springer, Heidelberg (2010)
5. van der Aalst, W.: Process Mining. Springer, Heidelberg (2016)
6. Mendling, J., Berente, N., Seidel, S., Grisold, T.: Pluralism and pragmatism in the information systems field: the case of research on business processes and organizational routines. Data Base Adv. Inf. Syst. **52** (2021)
7. Niehaves, B., Poeppelbuss, J., Plattfaut, R., Becker, J.: BPM capability development – a matter of contingencies. Bus. Process. Manag. J. **20**, 90–106 (2014)

8. vom Brocke, J., Maedche, A.: The DSR grid: six core dimensions for effectively planning and communicating design science research projects. Electron. Mark. **29**, 379–385 (2019)

9. Frank, U., Strecker, S., Fettke, P., vom Brocke, J., Becker, J., Sinz, E.: The research field "modeling business information systems": current challenges and elements of a future research Agenda. Bus. Inf. Syst. Eng. **6**, 39–43 (2014)

10. vom Brocke, J., Mendling, J., Rosemann, M. (eds.): Business Process Management Cases, vol. 2: Digital Transformation – Strategy, Processes and Execution. Springer, Heidelberg (2021)

11. Harmon, P.: Business Process Change: A Guide for Business Managers and BPM and Six Sigma Professionals. Elsevier/Morgan Kaufmann Publishers, Amsterdam/Boston (2007)

12. Kettinger, W.J., Grover, V.: Toward a theory of business process change management. J. Manag. Inf. Syst. **12**, 9–30 (1995)

13. Andreasen, P.H., Gammelgaard, B.: Change within purchasing and supply management organisations – assessing the claims from maturity models. J. Purch. Supply Manag. **24**, 151–163 (2018)

14. Andersen, K.V., Henriksen, H.Z.: E-government maturity models: extension of the Layne and Lee model. Gov. Inf. Q. **23**, 236–248 (2006)

15. vom Brocke, J., Zelt, S., Schmiedel, T.: On the role of context in business process management. IJIM. **36**, 486–495 (2016)

Bridging the Gap: An Evaluation of Business Process Management Education and Industry Expectations – The Case of Poland

Piotr Senkus[1](✉) ⓘ, Justyna Berniak-Woźny[2] ⓘ, Renata Gabryelczyk[1] ⓘ,
Aneta Napieraj[3] ⓘ, Marta Podobińska-Staniec[3] ⓘ, Piotr Sliż[4] ⓘ,
and Marek Szelągowski[2] ⓘ

[1] Faculty of Economic Sciences, University of Warsaw, Warsaw, Poland
[2] Systems Research Institute of the Polish Academy of Sciences, Warsaw, Poland
[3] AGH University of Kraków, Kraków, Poland
[4] Faculty of Management, University of Gdańsk, Gdańsk, Poland

Abstract. The fast-growing job market for Business Process Management (BPM) specialists and the need to teach the basics of BPM to students of technical, economic, and even medical faculties pose unique challenges to academic institutions that want to equip students with appropriate skills and knowledge. This article uses a two-pronged methodological approach to assess the compatibility between university BPM education programs and labor market requirements. First, web scraping techniques were used to analyze job offers in the BPM area, identifying essential skills and areas of knowledge required by employers. Second, a study of BPM subject syllabuses was conducted to gain insight into perceived gaps in current educational practices. The above research was preceded by a literature review on the education and skills required to implement and use BPM. The study results indicate a significant discrepancy between university BPM programs and the labor market requirements. The article proposes to significantly reduce this gap by, in cooperation with businesses, standardizing BPM courses at universities, starting with the standardization of teaching business process modeling following BPMN and the introduction of nationwide certification of modeling skills in BPMN. These changes will provide students with a more consistent, market-oriented learning experience, potentially increasing their job readiness and employability. The study underscores the urgency for academia to proactively adapt to labor market trends and contributes to the ongoing discourse on the future trajectory of BPM education.

Keywords: Business process management · BPM · education · curriculum · skills · competencies · labor market

1 Introduction

Business Process Management (BPM) has been helping organizations build and maintain a competitive advantage in the market for several decades by constantly adopting and implementing the best management practices, strategies, and technologies [8]. Focusing

J. Köpke et al. (Eds.): BPM 2023, LNBIP 491, pp. 250–262, 2023.
https://doi.org/10.1007/978-3-031-43433-4_18

on well-defined and optimized processes across the entire value chain is the foundation of an organization's success. However, maintaining this success requires creating value by effectively managing, organizing, communicating, and transforming business processes. These efforts require a range of skills and competencies that many organizations find difficult to acquire and develop. Previous research shows that organizations need employees with knowledge of BPM methodologies and notations [15], and the lack of competent employees in this area is a significant organizational problem [13, 21]. To meet those needs, numerous organizations introduce training and certification programs. For example, in 2008, ABPMP introduced a general model curriculum for BPM specialists, which was the first attempt to define comprehensive educational requirements for BPM practice. But talent development in this area cannot occur through ad hoc development interventions but must start at the Higher Education Institutions level.

Process management has been over 100 years, but it did not appear in university teaching until the late 1990s. Its main weakness is the lack of standardized programs to support universities and academic teachers in developing programs that meet the needs of a rapidly changing market driven by digital transformation, entailing process transformation of the organization [10, 16]. BPM is offered mainly in computer science and business management study programs [7]. Still, it also appears in other fields of study, e.g., in the area of health care [9]. In the literature, we can find many inspiring programs [1], courses [5] and activities [3] proposing to refer to all or selected topics, such as Initial BPM concepts, Process modeling, Process analysis, Process design and improvement, Process implementation Management, BPM strategy, and BPM technology and architecture. The article presents research aimed at answering the following research questions: RQ1 What essential skills and knowledge areas are currently being emphasized in university BPM curricula? RQ 2 What skills and knowledge areas are currently most in demand in the BPM job market? RQ 3 How well do existing BPM curricula align with the demands of the BPM job market? Leading to the main conclusion and recommendations: What changes could be made to BPM curricula to better align with job market requirements?

The article is structured as follows: The second part presents the research methodology. Part three provides an overview of the literature on education and the shortcomings of BPM education. The following section presents a study of current job offers for positions related to BPM. Part 5 gives an overview of the syllabuses of Polish universities on subjects associated with BPM and provides the reference syllabus for the BPM course. The results of comparing requirements according to the literature and analysis of job and educational offers are presented in part 6, Discussion and Conclusions.

2 Methodology

The research methodology was designed to provide a comprehensive evaluation of the current state of BPM education in Poland and its alignment with industry expectations. The methodology consisted of several stages. The first stage involved a narrative literature review, which served as the foundation for understanding the existing body of knowledge on BPM education. The review incorporated relevant scholarly articles, books, and reports that discussed the current BPM education approach and its relevance to industry needs and challenges. The second stage to gain insights into the current market demand for BPM skills and competencies, a job offer analysis was performed using the Pracuj.pl portal, the leading job portal in Poland. This analysis utilized web scraping techniques (WebHarvy software) for analyzing job postings. Next, to understand the current state of BPM education, the existing BPM syllabi from leading Polish universities were analyzed. This analysis followed the framework provided by [14] for evaluating BPM curricula, considering the courses and topics covered in the syllabi. Finally, to ensure a comprehensive understanding of the market, a customer segment analysis was conducted, and user personas were developed. This analysis followed the approach suggested by [2] to identify and understand the diverse needs and expectations of potential students interested in BPM education.

3   Literature Review

Research shows that the required competencies vary depending on the role in implementing or using BPM. This means we have different expected types and profiles of BPM specialists [1]. Based on industry best practices, four essential business process (BP) positions can be defined: (1) BP director; (2) BP consultant; (3) BP architect; and (4) BP analyst. Each of these positions is associated with varied tasks and roles, which leads to the conclusion that BPM is a border-crossing field that requires interdisciplinary sets of competencies, from technical competencies to business and systems competencies [11]. Thus we have mostly the competency frameworks dedicated to a specific role, among others, the one created by [20], who divided the competencies of a process analyst into five layers: Fundamental – business analysis, process and holistic thinking, customer orientation. Interpersonal – facilitation and leadership, communication, credibility, Organizational – understanding the strategy and links between functional departments, Process approach – BPM support, risk assessment modeling, process improvement and Technical – service-oriented architecture, ERP system, user interface design. A similar framework was developed by [12], but this time it is broader and includes competencies for the process analyst, process architect, and process professional. These frameworks include process modeling and redesign, performance measurement, workflows, governance and compliance management, BPM maturity, manual and procedure writing, surveys, stakeholder relationship management, project management, enterprise architecture, and Lean or Six Sigma certification. An attractive competency model was created by [19], who defined seven levels of BPM competencies among each employee in the organization (without reference to a specific role).

All these frameworks suggest that business process management competencies cover core business process management, strategic alignment, organizational goal setting, governance, documentation, training, and systems thinking [4].

4   Mapping BPM Components to Job Requirements

In this section, we meticulously examined a comprehensive dataset comprising over 90,000 current job postings for all types of positions (from leader to staff and various employment relationships). Due to the context of the study, it was decided to select specific job offers that aspire to the research area associated with BPM positions. The dataset was obtained from pracuj.pl, widely recognized as one of Poland's largest and most reputable job portals. Notably, this database has been previously cited and utilized in scholarly literature, addressing similar research inquiries [18, 23]. During the screening process, which involved keyword analysis focusing on the term "process" and its Polish equivalent, a total of 644 current job offers to align with the research objectives were identified for in-depth analysis. It is crucial to emphasize that all the selected job offers were directly associated with processes, although not exclusively limited to BPM. Among these job offers, there was a significant emphasis on positions related to production processes. This pattern indicates a substantial demand for professionals proficient in technological processes such as logistics and quality management and specialized expertise in executing specific processes like painting, galvanization, and tempering.

Regarding job classifications, it is notable that most of the analyzed job offers (88.50%) were oriented toward specialist and expert roles, indicating a demand for individuals with niche expertise in their respective fields. Additionally, a smaller proportion of the job offers (4.95%) targeted managerial positions across various hierarchical levels, encompassing lower, middle, and upper management. A relatively minor percentage of the job offers (5.27%) were intended for physical labor roles, while the remaining job offers were specifically for internships. Regarding employment terms, 99% of the job offers were for full-time positions. Most of the job offers were for permanent contracts (89.90%). Some job offers explicitly indicate the search for multiple candidates, which may signify high demand for specific positions.

As a result of the analysis, a set of soft and hard competencies was diagnosed, characteristic of positions related to the business processes. When analyzing the frequency of appearance of soft skills offered in the selected set, the most popular feature should be the ability to deal with problems and solve them efficiently. Next in the rank of importance were accessibility and communicativeness.

In the surveyed location, what could be important for a young employee that most posted job offers require short experience. Usually, it is a period of 1 to 3 years or even no indication. The condition set in this way reflects the novelty of this type of work area, which should be considered authentic because conscious business process management in Polish companies is in its early stages [17].

The most common requirement in the content of offers (Fig. 1) was the knowledge of foreign languages, especially English (63%). Employers also mentioned German as the second advantage. Higher education (50%) is placed among the most common requirements. Here, rather technical engineering is preferred.

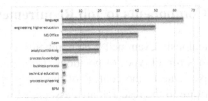

Fig. 1. List of employers' requirements in the surveyed set of job offers.

The requirements for higher education in the analyzed set of offers most often concern production engineering, but they differ slightly depending on the industry. These include chemical, medical, materials, mechanical, and process engineering, which often occur together. Fluent computer and MS Office skills are usually required (40%). A similar percentage of occurrence among the analyzed offers, i.e., 20%, was achieved by analytical thinking, practical knowledge, and the ability to use tools and methods in the field of Lean. Skills in the field of quality management methods and tools are also indicated. Here, the most frequently mentioned are PFMEA, APQP or SPC. Requirements for knowledge of BPM occurred in only a few offers.

When analyzing in detail job positions with process, product, or production engineer in the title, 396 offers were selected from the entire surveyed set (Table 1). These offers were the subject to further analysis.

Table 1. The frequency of occurrence of selected jobs.

Class	Number	Cumulative number	Percent	Cumulative percent
Process Engineer	217	217	54.79	54.79
Production/Product Engineer	179	396	45.20	100
Missing data	0	396	0.00	100

It was noticed that many of these advertisements included requirements for knowledge of Lean Manufacturing methods or quality management, such as Kaizen, 5S, Just in Time, Kanban, TPM, DMAIC, FMEA, Control Plan, 8D, 5Why, Ishikawa diagram, Pareto, PDCA. An essential clue in the detailed analysis was that mechanics was the preferred education in 93 advertisements, automatics seemed as many as 62 times for the entire surveyed set, robotics 32 times, and production engineering 38 times.

Comparing soft and hard competencies selected based on the analysis of advertisements posted on the Pracuj.pl portal with those based on the literature review, it should be emphasized that the soft competencies indicated correspond to the group of features correlated with interpersonal competencies. On the other hand, moving to the essential organizational functions, approach, or technical process, one should notice a big difference between the state that should be presented and the existing one. There is no clear emphasis on purely process competencies in the analyzed advertisements. In addition, this discrepancy reappears when we compare the competency model created by [19].

In the analyzed advertisements, no distinction was made regarding the level of skills in modeling processes. Therefore, it is difficult to indicate advertisements developed for a beginner employee in the field of processes or an employee with expert knowledge. The result of this observation should be a study of whether competent business process leaders are being sought in the Polish market. It is about leaders who will know the entire value chain of the company, professionally model the process, and be able to define change and implement it in processes. The juxtaposition of the literature with the Polish market indicates an undeveloped work area that requires clarification and, thus, the supply of competent "process workers" to the labor market.

5   Results and Findings – The Reference Syllabus Proposal

Organizations in diverse industries strive for high-quality products, regulatory compliance, and operational excellence. Achieving these goals requires a systematic approach to process documentation, ensuring well-defined, consistent, and transparent operations. International standards like ISO 9001, ISO 22000, ISO 14001, ISO 27001, ISO 45001, ISO 20000–1 and ISO 13485 provide frameworks for organizations to document their processes and achieve desired outcomes.

In addition to existing standards, future challenges in BPM involve the use of RPA and AI. Organizations must explore how to effectively document and integrate RPA and AI tools within their BPM frameworks. This requires understanding the potential benefits, addressing ethical considerations, and adapting existing process documentation practices to embrace these emerging technologies.

5.1   Business Process Management Syllabi at Polish Universities Analysis

The analysis of existing BPM syllabi at Polish universities reveals a significant discrepancy between academic offerings and the labor market requirements. The educational offerings are neither in line with business requirements nor the recognized competence models presented in the literature (see Table 3). The educational programs are designed for an average graduate, which caters to no one in particular. Despite the vast number of courses offered by leading universities in Poland, BPM is taught in only four to five courses. For instance, the University of Warsaw provides over 300 programs of 1st, 2nd, and long-cycle – uniform Master's studies, yet BPM is scarcely represented. Similar trends are observed at the Jagiellonian University, Warsaw University of Technology, Adam Mickiewicz University in Poznań, AGH University of Science and Technology in Kraków, Gdańsk University of Technology, Medical University of Gdańsk, Wrocław University of Science and Technology, and Medical University of Łódź (Table 2).

The key role competencies was identified through a comprehensive analysis of the course content and the explicit learning outcomes in the syllabi. The competency from job offers was derived from the skills and qualifications listed in the job descriptions. For instance, process modeling was a common requirement in many job offers. The assessment score represents the degree of alignment between the competencies taught in the courses and those demanded in the job market. The score was determined through a comparative analysis of the course content and job requirements.

Table 2. List of the analyzed universities.

University	Number of Courses	Number of BPM Courses
University of Warsaw	about 300	3
Jagiellonian University	87	8
Warsaw University of Technology	32	5
Adam Mickiewicz University in Poznań	228	3
AGH University of Science and Technology in Kraków	23	8
Gdansk University of Technology	41	6
Medical University of Gdansk	17	2
Wroclaw University of Science and Technology	50*	6

5.2 Analysis of the Requirements of Potential Users

The user personas presented in this study were identified based on the job offers analyzed. We examined the skills, qualifications, and job roles mentioned in these offers and used this information to create representative personas. These personas, such as the healthcare professional, the manufacturing engineer, and the recent computer science graduate, were designed to represent potential students' diverse backgrounds and career aspirations. These personas provide a realistic representation of the job market and can help design a BPM curriculum that caters to the needs of various industry sectors.

The dynamic nature of BPM development also requires a flexible approach to course development. To ensure industry relevance, it is essential to comprehend the specific requirements and expectations of prospective students from different industries. The result of user personas is an effective method for achieving this comprehension. User personas offer valuable insight into prospective students' motivations, objectives, and backgrounds. As a healthcare professional, Rachel desires to comprehend how BPM can enhance patient treatment. A manufacturing engineer, Sam wishes to learn how BPM can optimize supply chain processes. A recent Computer Science graduate, Emily is curious about digital technologies' function in business process management (BPM). Understanding these nuances is essential for designing a BPM curriculum that meets the requirements of the students and enhances their employability. Table 4 presents the list of the user personas – potential students of the BPM course.

Table 3. Evaluation of the courses related to BPM

Role	Key Role Competency	Corresponding Competency from job offers	BPM Role Preparation Assessment*
Process Owner	Responsibility for the whole cross-functional process	Process knowledge	4
	Documenting the process	Process knowledge	4
	Standardization within individual branches	Organization skills	2
	Authorizing process variants	Independence	2
	Approving process improvements	Troubleshooting ability	2
	Ensuring that changes do not negatively affect other processes and workers	Working under pressure	2
BPM Manager	Alignment between business needs and processes	Process knowledge	3
	Leadership and motivation	Communication skills	2
	Delivering process modeling and improvement to internal customers	Process modeling	4
	Serving as a connection between business and information technologies	-	2
Process Analyst	Process modeling and writing related documentation	Technical education	3,5
	Simulation	Technical education	3
	Ensuring alignment between tools, supporting performance measurement system, internal customers, and improvement proposals	Organization skills	3

(*continued*)

Table 3. (*continued*)

Role	Key Role Competency	Corresponding Competency from job offers	BPM Role Preparation Assessment*
	Serving as a connection between business and information technologies	Technical education	2
Industrial Engineer	Professional knowledge of change management, statistical methods, and process design	Process engineering	2
	Professional knowledge of change management, statistical methods		2
	Training other black and green belts		2
	Sponsoring improvement projects and taking over results		2
BPM Excellence Manager	Responsibility for the BPM program, individual projects, and continuous development	Organization skills	3
	Coordination of all process work in the organization	Process knowledge	3
	Monitoring process efficiency	Analytical thinking	2
	Responsibility for the BPM program, individual projects, and continuous development		3
	Coordination of all process work in the organization		3

Table 3. (*continued*)

Role	Key Role Competency	Corresponding Competency from job offers	BPM Role Preparation Assessment*
	Providing support for process change management		2
Other required competencies for future roles in BPM	Creativity		2
	Specific industry knowledge		2
	Digital technologies AI utilization in BPM activities (AI, RPA)		2
	Problem-solving		2
	Knowledge of modern management concepts		2

* Overall courses characteristics – Hours – 15–30 (22 average) Target studentsmainly economic and management faculties, rarely IT and production faculties ** Dominant value.
Source: Own analysis, inspired by [5]

5.3 Reference Syllabus Proposal

Based on the previous analysis, syllabus of the course entitled "Introduction to Business Process Management: From Basics to Advanced Applications" has been proposed. It gives the students a thorough comprehension of BPM and its implementation in diverse sectors. The objective of the course is to narrow the divide between scholarly expertise and industry demands through a concentration on essential proficiencies necessary in the BPM sector, such as analytical thinking, effective communication, innovative thinking, and adeptness in handling stressful situations. The syllabus has been designed comprehensively to facilitate a comprehensive comprehension of BPM. The course begins by introducing fundamental concepts and theories and subsequently advances to more sophisticated applications, such as routine dynamics, lean management, and quality management. The approach also encompasses a robust emphasis on digital technologies within the realm of BPM, which comprises the utilization of future trends like RPA and AI (Table 4).

Compared to existing curricula and the ACM/AIS reference curriculum for Information Systems (IS 2020), proposed curriculum emphasizes a more industry-specific approach to BPM education. It incorporates vital BPM components identified from job market analysis, such as RPA and Process Automation, which are currently underrepresented in existing curricula. However, we acknowledge that a detailed comparison of our proposed curriculum with existing curricula and the IS 2020 reference curriculum is beyond the scope of this study and will be subject to further research. This will allow us to refine our curriculum proposal and ensure its relevance and effectiveness in preparing students for the BPM job market.

Table 4. Course syllabus proposal

Course Description: This course aims to introduce students to the fundamentals of Business Process Management (BPM) and equip them with the knowledge and skills to understand and manage business processes in various industries. The course will cover basic concepts, methods, and theories of BPM, routine dynamics, lean management, and quality management. It will also focus on developing key competencies required in the BPM industry, including problem-solving, communication, creativity, and the ability to work under pressure

Course Objectives:
- Understand the basic concepts and theories of BPM, Routine Dynamics, Lean Management, and Quality Management
- Develop essential soft skills and competencies required in the BPM industry
- Understand the role of digital technologies in managing business processes
- Apply the knowledge and skills acquired in the course to real-world business scenarios
  1. Introduction to BPM (3 h) – A) Definition and importance of BPM; B) Basic concepts and theories of BPM;
  2. Routine Dynamics (3 h) – A) Understanding routine dynamics in business; B) Role of routine dynamics in process change;
  3. Lean Management and Quality Management (6 h) – A) Striving to simplify processes and increase their transparency; B) Liquidation of processes that do not bring added value for the customer; C) Emphasis on flawless performance and treating errors as opportunities for improvement; D) Constant search for perfection and use of tools to ensure proper measurement and quality assurance; E) Control, information, and problem-solving at the lowest possible level of the organization;
  4. Soft Skills and Competencies (6 h) – A) Problem-solving and decision-making in relation to processes; B) Communication and teamwork in process improvement; C) Creativity and innovation in process improvement; D) Working under pressure and stress management;
  5. Digital Technologies in BPM (3 h) – A) Role of digital technologies in BPM; B) Impact of digital transformation on business processes; c) Introduction to RPA and AI in BPM;
  6. Industry-Specific BPM (12 h) – A) Application of BPM in various industries (IT, Banking, Healthcare, Manufacturing); B) Case studies on how BPM is implemented in these industries;
  7. BPM Roles and Competencies (3 h) – A) Overview of key BPM roles and their competencies; B) Case studies on how these roles are implemented in different industries;
  8. Course Project (6 h) - Students will be required to design a business process for a hypothetical business in their chosen industry and manage its dynamics using the knowledge and skills acquired in the course

6 Discussions and Conclusions

As the study shows, many competency models are available in the literature that could be used when creating job offers for positions requiring competencies in BPM. Unfortunately, the analysis of job offers clearly shows that they are made without the use of competency models and a broader view of the benefits of a holistic approach to BPM implementation. Undoubtedly, the result of the analysis demonstrates a significant predominance of job offers for technical positions, where work efficiency is easy to plan and measure. In addition, it does not require knowledge about the benefits of holistic implementing BPM beyond simply understanding increasing productivity, reducing costs, or improving quality.

A comparison of Polish universities' business expectations and educational offers shows significant differences between them. In general, the educational offer of Polish universities is neither in line with business requirements nor with the recognized competence models presented in the literature. The authors of the study believe that the best direction to heal this situation would be to use the experience and achievements of project management in teaching BPM. This would require e.g., basing the scope of teaching on standards recognized in business and proposing to confirm the skills of students (as well as employees of companies dealing with BPM) based on independent certification paths. Such programs are great for project management (PMI, Prince) and even validation of computer skills (European Computer Driving License, ECDL).

In our conclusions, we acknowledge that while business expectations may not always be adequate for the current level of BPM development, they provide a realistic snapshot of the current market demands. Educational institutions must be aware of these demands to prepare students effectively for the job market. Education should not merely mirror business expectations but should also aim to shape them by equipping students with the knowledge and skills to drive innovation and change in business practices.

The presented study has two significant limitations. Firstly, it concerns only offers on the Polish labor market and additionally offers from one quarter of 2023. The second limitation is the analysis of syllabuses of only ten leading Polish universities.

Future research will focus on the analysis of selected study programs and selected subjects and then the preparation of recommendations agreed upon within the university, or perhaps a Conference of Rectors of Academic Schools in Poland, regarding education in specific groups of skills in the BPM area (e.g., business process modeling, business process implementation, RPA). The second direction of future research will be its extension to other Europe countries and beyond.

References

1. Antonucci, Y.L.: Business process management curriculum. In: vom Brocke, J., Rosemann, M. (eds.) Handbook on Business Process Management 2. IHIS, pp. 547–572. Springer, Heidelberg (2015). https://doi.org/10.1007/978-3-642-45103-4_23
2. Cooper, A., et al.: About Face: The Essentials of Interaction Design. John Wiley and Sons, Indianapolis (2014)
3. Ettl, F., et al.: Updated teaching techniques improve CPR performance measures: a cluster randomized, controlled trial. Resuscitation **82**(6), 730–735 (2011). https://doi.org/10.1016/j.resuscitation.2011.02.005
4. Gabryelczyk, R. et al.: Motivations to adopt BPM in view of digital transformation. Inf. Syst. Manage. 1–17 (2022). https://doi.org/10.1080/10580530.2022.2163324
5. Grisold, T. et al.: Managing process dynamics in a digital world: integrating business process management and routine dynamics in IS curricula. CAIS **51**, 637–656 (2022). https://doi.org/10.17705/1CAIS.05127
6. Harmon, P.: Business Process Change: A Guide For Business Managers and BPM and Six Sigma Professionals. Elsevier/Morgan Kaufmann Publishers, Amsterdam (2010)

7. Koch, J., et al.: Theory and practice - what, with what and how is business process management taught at German Universities? In: Di Ciccio, C., et al. (eds.) Business Process Management. pp. 34–39. Springer International Publishing, Cham (2022). https://doi.org/10.1007/978-3-031-16103-2_4

8. Kohlbacher, M., Gruenwald, S.: Process ownership, process performance measurement and firm performance. Int. J. Productivity Perf. Mgmt. 60(7), 709–720 (2011). https://doi.org/10.1108/17410401111167799

9. Mang, H. et al.: Medical process management - an innovative master of science program addressing the challenges faced by health care systems. In: Dössel, O., Schlegel, W.C. (eds.) World Congress on Medical Physics and Biomedical Engineering, 7–12 September 2009, Munich, Germany, pp. 366–367 Springer, Heidelberg (2009). https://doi.org/10.1007/978-3-642-03893-8_105

10. Marjanovic, O., Bandara, W.: The current state of bpm education in Australia: teaching and research challenges. In: zur Muehlen, M., Su, J. (eds.) BPM 2010. LNBIP, vol. 66, pp. 775–789. Springer, Heidelberg (2011). https://doi.org/10.1007/978-3-642-20511-8_69

11. Müller, O., et al.: Towards a typology of business process management professionals: identifying patterns of competences through latent semantic analysis. Enterp. Inf. Syst. 10(1), 50–80 (2016). https://doi.org/10.1080/17517575.2014.923514

12. Panagacos, T.: The ultimate guide to business process management: everything you need to know and how to apply it to your organization. CreateSpace Independent Publishing Platform, s.l. (2012)

13. Pridmore, J., Godin, J.: Business process management and digital transformation in higher education. IIS (2021). https://doi.org/10.48009/4_iis_2021_180-190

14. Recker, J.: Scientific Research in Information Systems: A Beginner's Guide. Springer International Publishing, Cham (2013)

15. Sarvepalli, A., Godin, J.: Business process management in the classroom. J. Cases Inf. Technol. 19(2), 17–28 (2017). https://doi.org/10.4018/JCIT.2017040102

16. Seethamraju, R.: Business process management: a missing link in business education. Bus. Process. Manag. J. 18(3), 532–547 (2012). https://doi.org/10.1108/14637151211232696

17. Sliż, P.: Organizacja procesowo-projektowa: istota, modelowanie, pomiar dojrzałości. Difin, Warszawa (2021)

18. Sliż, P.: Robotization of business processes and the future of the labor market in poland -preliminary research. Organizacja i kierowanie. 185(2), 67–79 (2019)

19. Smith, H., Fingar, P.: Business Process Management: The Third Wave. Meghan-Kiffer Press, Tampa (2007)

20. Sonteya, T., Seymour, L.F.: Towards an understanding of the business process analyst: an analysis of competencies. J. Inf. Technol. Educ. Res. 11, 1 (2012)

21. Szelągowski, M., et al.: The importance of certification of business process modelling competencies. Presented at the 37th International Business Information Management Association Conference (IBIMA), Cordoba, Spain 31.05 (2021)

22. Vom Brocke, J.V., et al.: On the role of context in business process management. Int. J. Inf. Manage. 36(3), 486–495 (2016). https://doi.org/10.1016/j.ijinfomgt.2015.10.002

23. Wiechetek, Ł., et al.: Business process management in higher education. The case of students of logistics. PZ 15(4(71)), 146–164 (2017). https://doi.org/10.7172/1644-9584.71.10

Author Index

Printed in the United States
by Baker & Taylor Publisher Services